THE RARE METALS WAR

Guillaume Pitron, who was born in 1980, is a French award-winning journalist and documentary-maker for France's leading television channels. His work focuses on commodities and on the economic, political, and environmental issues associated with their use. *The Rare Metals War* is his first book, and has been translated into eight languages. Guillaume Pitron holds a master's degree in international law from the University of Georgetown (Washington, DC), and is a TEDx speaker. More information at www.guillaumepitron.com.

Bianca Jacobsohn is a South African/French translator and conference interpreter who specialises in energy, strategic metals, and diplomacy. More information at www.biancajacobsohn.com.

THE RARE METALS WAR

the dark side of clean energy and digital technologies

GUILLAUME PITRON

SCRIBE

Melbourne | London | Minneapolis

Scribe Publications
18–20 Edward St, Brunswick, Victoria 3056, Australia
2 John St, Clerkenwell, London, WC1N 2ES, United Kingdom
3754 Pleasant Ave, Suite 100, Minneapolis, Minnesota 55409, USA

First published in French as *La guerre des métaux rares* by Les Liens qui Libèrent
in 2018

First published by Scribe 2020
This updated edition published 2024

Typeset in Portrait by the publishers

Printed and bound in the UK by CPI Group (UK) Ltd, Croydon CR0 4YY

Scribe is committed to the sustainable use of natural resources and the use of
paper products made responsibly from those resources.

978 1 761380 95 2 (Australian edition)
978 1 914484 96 4 (UK edition)
978 1 957363 90 5 (US edition)
978 1 925938 60 9 (ebook)

Catalogue records for this book are available from the National Library of
Australia and the British Library.

scribepublications.com.au
scribepublications.co.uk
scribepublications.com

To my father,
to my mother

CONTENTS

Foreword

by Hubert Védrine

French minister of foreign affairs under President Jacques Chirac,
and secretary general and diplomatic adviser to French president
François Mitterrand

IN AN INCISIVE AND TROUBLING ACCOUNT, GUILLAUME
Pitron sounds the alarm on a serious geopolitical problem: the
world's growing reliance on rare metals for its digital development
in information and communication technologies. This includes
the manufacture of devices such as mobile telephones, not to
mention the much-lauded electric and/or hybrid car, which
requires twice as many rare metals as the humble internal-
combustion engine vehicle.

There is nothing untoward about these thirty or so rare
metals bearing perfectly civilised Latin names like 'promethium'.
They are found in minute proportions in more abundant
metals, making their extraction and refinement expensive and
difficult. The first problem is that most of these resources are in

the hands of China — an advantage it is naturally tempted to exploit. Other countries with such underground resources have for various reasons abandoned their mining operations, which largely gives China a global monopoly and Beijing the title of the 'New Rare Metals Master'.

Pitron illustrates the perils of this dependence with numerous case studies — ranging from super magnets to long-range missiles — where the West has acted inconsistently or entirely without foresight. The solution seems obvious: reopen rare metal production in the United States, Brazil, Russia, South Africa, Thailand, Turkey, and even in the 'dormant mining giant' of France.

Enter the next predicament: mining these rare minerals is anything but clean! Says Pitron, 'Green energies and resources harbour a dark secret.' And he's quite right: extracting and refining rare metals is highly polluting, and recycling them has proved a disappointment. We are therefore faced with the paradox that the latest and greatest technology (and supposedly the greenest to halt the ecological countdown) relies mostly on 'dirty' metals. Thus, information and communication technologies actually produce 50 per cent more greenhouse gases than air transport! It's an especially vicious circle.

How do we overcome the contradiction?

We need to revive the mining of rare earths and of mineral resources internationally (potentially reviving tensions between governments and mining companies), but in an environmentally sound way, using the latest financing, innovation, and other economic and technological means. According to Pitron, more and more consumers around the world would be willing to foot the bill.

The author ends his thesis on a positive note by giving examples of the 'sudden wake-up call taking place in the rare metals industry'.

The ecological transition of our economic activities is critical, not just for saving the planet, but for preserving *life* on the planet — including human life. We can expect hundreds more such challenges to overcome, difficult decisions to be made, scientific breakthroughs to reach, and opinions to support or enlist if we are to accelerate this transition. Meanwhile, the clock is ticking.

Through the focus of his investigation, Guillaume Pitron alerts us to an issue that is vital yet inadequately considered.

Preface

IN THE YEARS SINCE *THE RARE METALS WAR* WAS PUBLISHED in 2018, the subject of rare, critical, and strategic metals has become a global talking point. By spring 2023, it was inescapably clear that the book needed to be entirely updated. This updated version is what you now hold in your hands.

In 2023 the European Parliament announced that carmakers may sell only 100 per cent electric vehicles in Europe from 2035. This has accelerated our need for batteries, and triggered in Europe and around the world the eruption of 'gigafactories' — the new symbols of humanity's green shift.

But how do we manufacture such technology without mining raw materials? According to one study, meeting the demand for electric vehicle batteries alone by 2035 will require opening almost 400 new mines around the world (ninety-seven for natural graphite, seventy-four for lithium, and seventy-two for nickel).[1] The predicted boom in our metal requirements is awakening the United States and Europe from its protracted slumber, as states and carmakers form procurement partnerships left and right with mining countries (such as Chile, Indonesia, Ghana, and Canada), now basking in the glow of a new-found prestige.

The ecological and economic risks inherent in mining operations are compelling a growing number of militant environmentalists and politicians to take a stance on controversial debates, including the partial reshoring of metal production to the West, and the extraction of polymetallic nodules from the seafloor.

A final, more geopolitical context has further underscored the importance of the themes covered in this book. The Covid-19 and Russian gas crises opened our eyes, especially in Europe, to our reliance on countries whose strategic agenda differs from the West's for medical equipment (masks) and resources (gas). Two years went by between the first lockdown and Russia's invasion of Ukraine ... yet it was enough to completely overturn our perception of globalisation. While the 'invisible hand of the market' should have guaranteed us unfettered access to resources, globalisation is now seen as a force of economic dependence and strategic vulnerability.

Another reason for publishing this updated edition is the book's continued sales success. The original French-language work has sold over 100,000 copies, and has been, or is in the process of being, translated into eleven languages in some fifteen countries. And there is little reason to believe it will lose momentum any time soon, as future developments increasingly give pride of place to the 'new oil' that is mining resources. My goal is that, with each new edition, *The Rare Metals War* will remain an authoritative source on the subject.

I must at this point reemphasise my immutable credo:

investigating and writing about the 'dark side' of the energy transition does not, in any way, mean that we should stop decarbonising the way we live our lives. Quite the opposite! 'We must choose the right target', confirms Olivier Vidal, researcher at the French National Centre for Scientific Research (CNRS) and mineral resources specialist. 'This transition will come at a cost until 2050; after that we will reap the benefits for the next 500 years.'[2] It bears repeating: I write and investigate not for the energy transition to be constrained, but rather for it to be accelerated, and its negative effects mitigated. This is a challenge that will involve a revival of mining — whether applauded or abhorred — that is both large scale and sustainable.

The energy transition is truly the only way forward. Yet scratching the surface of rare metals opens a well of questions, and just as many challenging prospects.

How will we share the costs and benefits of the energy transition equitably as more and more developing countries will have to bear the environmental burden of extracting rare earths so that we in the West can drive around in 'zero-emission' electric SUVs?[3]

How can we reconcile democracy with the environment, when public opposition could derail plans in Western countries to reopen lithium and rare earth mines for the purposes of manufacturing green technology? And how far are we willing to go to change our behaviour as consumers, while the circular economy is slow to take root?[4] It takes 20 tonnes of material extracted from the earth every year to satisfy the needs of just one European.[5] This is incompatible with sustainable lifestyles.

These are difficult questions to answer; however, recent developments reveal both the common sense and illogical nature of the human spirit in the face of such colossal challenges. Some of these developments are of particular concern.

Start with the project put forward in June 2023 by the Norwegian government to authorise mining for metals in the seabeds of its exclusive economic zone. What will be the ecological cost of such operations? With no definitive answer at hand, we can start by looking at the heavy machinery that has already left the shopfloor of British manufacturer SMD (Soil Machine Dynamics), and which could soon be ploughing the seabed. Take the bulk cutter — a mobile excavator weighing 300 tonnes, measuring 15 metres in length, and fitted with two 800 horsepower engines, which was originally slated for deep-sea mining off the coast of Papua New Guinea.[6] Had the 2015 United Nations Climate Change Conference (COP 21) warned us then that weaning ourselves off petrol would mean today discussing the use of such a beast of a bulldozer thousands of metres underwater, would we have celebrated the Paris Accord so jubilantly?

In 2021, an announcement by Rolls-Royce also plunged us into uncertainty. In its drive to go electric, the British carmaker announced the launch at the end of 2023 of its Spectre (the 'Spirit of Ecstasy' redesigned) — a new electric super coupé with a curb weight of 2,975 kilograms. Including passengers, its gross vehicle mass could come within a hair's breadth of 3.5 tonnes. Beyond this limit, the Spectre is considered a large-goods vehicle rather than a car, requiring a Category C driving licence! That's why another carmaker — Bentley this time — has asked European authorities to raise the 3.5 tonne weight limit that future electric ultra-luxury vehicles could exceed.[7] Does this commercial strategy truly fit in with the spirit and intent of the 'green' transition?

Fortunately, two recent developments give us cause for hope:

The first is the adoption in 2022 of the European Corporate Sustainability Reporting Directive, to be phased in from 1 January 2024. Under the directive, certain companies are required to supplement their traditional accounting with an environmental and human-impact assessment of their corporate activities. This 'triple bottom-line' accounting that appraises social, environmental, as well as economic value will disclose the environmental and social gains and losses incurred by economic players — unlike conventional accounting methods that record only financial performance. It is without question a Copernican revolution, for it will make us aware that we never pay the true price for our consumer goods; that the price we pay is undervalued — rock-bottom, even — given the impacts of mining for raw materials on health, biodiversity, and soil fertility.

How will tax authorities enforce such mandatory requirements in the future? What will be the impact on our buying power? Already, triple bottom-line accounting has the makings of a potentially explosive issue, at a time when consumers are more determined than ever to revel in the benefits of the technologies surrounding them — including the 'greenest' of them — without having to put up with the slightest inconvenience.

Another cause for hope is the countless numbers of French, European, and American students who contact me for advice or insight for their theses and research papers on metals. The upsurge in interest in the geopolitics of lithium, nickel, and rare earths stands in stark opposition to the relative indifference shown by our political leaders on the subject, irrespective of their political stripes. We can count on these armies of young chemists, geologists, economists, diplomats, and other experts to buck this grim assessment as they stand poised to delve deeper into this relatively new topic, and to deliver new insights and possibilities.

Researching rare metals is as fascinating as it is strategic. It melds diverse disciplines such as geopolitics, economics, and materials science, and lays bare the very real challenge of preserving our common home — an imperative far from mind a century ago, when oil infiltrated our lifestyles. Rare metals compel us to be more discerning in how we think about the world around us.

They also tell the story of the past, while lifting the veil on the future, revealing both the sum of our shortcomings and the extent of our genius. These materials give us a glimpse of the exciting yet deeply complex 'green' world that lies ahead: a spyhole to better apprehend diplomatic powerplay, and the benefits of the latest technological breakthroughs. They allow us to scrutinise geological strata, the ocean floor, and even the treasures embedded in asteroids. Rare metals are a prism through which we may contemplate the universe.

Introduction

FOR 400,000 YEARS, HUMANS DEPENDED ON FIRE, CAPRICIOUS winds and currents, manpower and then horsepower to roam, build fortresses, and work the land. Energy was a rare and precious resource, movement was slow, economic growth sluggish. Progress came in fits and starts, and history tended to be made one slow step at a time.

Then, from the eighteenth century, humans used the steam engine to power their looms, propel locomotives, and float battleships to reign over the seas. Steam powered the first industrial revolution. This was the world's first energy transition, underpinned by the use of an indispensable fuel: a black stone called coal.

In the twentieth century, humans cast aside steam for another innovation: the petrol engine. This technology made vehicles, boats, and tanks more powerful, and paved the way for a new machine — the aeroplane — to conquer the skies. This second industrial revolution was also an energy transition, this time relying on the extraction of another resource: a rock oil called petroleum.

The disruptive effects of fossil fuels on the climate since the turn of the current century have driven humanity to develop new and supposedly cleaner and more efficient inventions — wind turbines, solar panels, electric batteries — that can connect to high-voltage ultra-performance grids. After the steam engine and the internal-combustion engine, these 'green' technologies have shifted us into a third energy and industrial revolution that is changing the world as we know it. Like its two predecessors, this revolution draws on a resource so vital that energy experts, techno-prophets, heads of state, and military strategists already refer to it as The Next Oil of the twenty-first century.

What resource are we talking about?

Most people don't have the slightest idea.

Changing the way we produce and therefore consume energy is humanity's next great adventure. Political leaders, Silicon Valley entrepreneurs, proponents of more moderate consumption, Pope Francis, and environmental groups have urged us to make this change, curbing global warming and saving ourselves from a second flood.[1] Never have empires, religions, and money been so aligned behind a single undertaking.[2] The proof of this — described by former French president François Hollande as the 'first universal agreement in our history' — is neither peace treaty, nor trade deal, nor financial regulation.[3] The Paris Agreement that was signed in 2015 following the twenty-first conference of parties to the United Nations Framework Convention on Climate Change (COP 21) is, in fact, an energy treaty.

The technologies we use every day might change, but our primary need for energy will not. Yet, faced with the question of what resource could possibly replace oil and coal as we embrace a new and greener world, no one really knows the answer. Our nineteenth-century ancestors knew the importance of coal, and the enlightened man on the street in the twentieth century was well aware of the need for oil. But today, in the twenty-first century, we are unaware that a more sustainable world is largely dependent on rock-borne substances called rare metals.

Humans have long mined the big names in primary metals: iron, gold, silver, copper, lead, aluminium. But from the 1970s, we turned our sights to the superb magnetic, catalytic, and optical properties of a cluster of lesser-known rare metals found in terrestrial rocks in infinitesimal amounts. Some of the members of this large family sport the most exotic names: rare earths, vanadium, germanium, platinoids, tungsten, antimony, beryllium, fluorine, rhenium, tantalum, niobium, to name but a few. Together, these rare metals form a coherent subset of some thirty raw materials with a shared characteristic: they are often associated with nature's most abundant metals.

As with all elements extracted from nature in Lilliputian quantities, rare metals are concentrates packed with remarkable properties. It is a long and painstaking process, for instance, to distil orange blossom essential oil, but the perfume and therapeutic powers of a single drop of this elixir continue to astound researchers.[4] Producing cocaine deep in the Colombian jungle is no easier feat, yet the psychotropic effects of just one gram of the powder can completely deregulate your central nervous system.[5]

The same applies to the rarest of the rare metals. Eight and a half tonnes of rock need to be purified to produce a kilogram of vanadium; sixteen tonnes for a kilogram of cerium; fifty tonnes for the equivalent in gallium; and a staggering 1,200 tonnes for one miserable kilogram of the rarest of the rare metals: lutetium.[6] (See the periodic table of elements in Appendix 1.) These effectively form the 'primary asset' of the Earth's crust: a concentration of atoms with outstanding properties, fine-tuned by billions of years of geological activity. Once processed industrially, a minute dose of these metals emits a magnetic field that makes it possible to generate more energy than the same quantity of coal or oil. And this is the key to 'green capitalism': the replacement of resources that emit billions of tonnes of carbon dioxide with resources that do not burn and therefore do not generate the slightest gram of it.

Less pollution, and at the same time a lot more energy. Tellingly, one of these elements was given the name 'promethium' by chemist Charles Coryell in the 1940s.[7] His wife, Grace Mary, suggested the name, based on the Greek myth of the Titan Prometheus, who was helped by the goddess Athena to break into the realm of the gods, Olympus, and steal the sacred fire ... to give to humanity.

The name says a great deal about the promethean power that we have acquired by harnessing rare metals. Like demigods, we have carved out a multitude of applications in two fundamental areas of the energy transition: supposedly 'green' technologies and digital technologies. Today, we are assured that the convergence of the two will create a better world. The first examples of this convergence (wind turbines, solar panels, and electric cars) are packed with rare metals to produce decarbonised energy that travels through high-performance electricity grids to enable power savings. Yet these grids are also driven by digital technology that is heavily dependent on these same metals. (See Appendix 13 for the main industrial applications of rare metals.)

Jeremy Rifkin, a leading US theorist of this energy transition and the resulting third industrial revolution, takes this a step further.[8] He writes that the crossover of green technologies and new technologies of information and communication already enables each of us to abundantly and inexpensively generate and share our own 'green' electricity. In other words, the mobile phones, tablets, and computers we use every day have become the key components of a more environmentally friendly economic model. Rifkin's prophecies are so compelling that he now counsels numerous heads of state, and is advising a region in the north of France on how best to implement its new-energy models.[9]

Recent history lends substance to his predictions: in the space of ten years, wind energy has increased seven-fold, and solar power by forty-four. In 2020, renewable energy already accounted for nearly 15 per cent of world final energy consumption.[10] Europe plans to increase its share to as much as 32 per cent by 2030.[11] Even technologies based on combustion engines use these metals to make vehicle and aircraft design lighter, more efficient, and therefore less fossil-fuel-intensive.

Enter the military, which is pursuing its own energy transition. Or *strategic* transition. While generals are unlikely to lose sleep over the carbon emissions of their arsenals, as oil reserves dwindle they will nevertheless have to consider the possibility of war without oil. Back in 2010, a highly influential American think tank instructed the US army to end its reliance on fossil fuels by 2040.[12] How will they do this? By using renewable energy and fleets of electric vehicles, including military vehicles.[13] Moving away from fossil fuels would solve the logistical conundrum of getting fuel to the front line.[14]

War is already colonising new — virtual — territories; digital infrastructures and communications networks must be protected, the ability to target the enemy's networks developed.[15] The armed conflict raging between Ukraine and Russia since 24 February 2022 highlights just how important digital technology has become for both ground and information warfare.[16]

Like army generals, we too are engaged in a transition to a connected world in which the way we use digital technology will replace certain resources with nothing but … thin air: clouds, intangible messaging, and online traffic instead of highway traffic. The digitalisation of the economy — we are assured — will drastically reduce our physical footprint on the living world. We stand only to gain from an energy and digital revolution: two technological forces marching hand in hand towards a better world.

Even the face of international relations is changing, as diplomats use rare metals to drive a geopolitical transition. Indeed, the rise of new non-carbon energy, say geopolitical experts, will upend the relationship between oil-producer states and oil-consumer states. It will enable the US to progressively shift its warships from the Straits of Hormuz and Malacca — today's vital oil-transit chokepoints — and rethink its partnership with the Gulf petro-powers. As for the European Union, less reliance on Russian, Qatari, and Saudi Arabian fossil-fuel imports will increase its member states' energy sovereignty.

For all these reasons, the energy transition promises to be positive — although implementing it will be no easy feat so long as we have not seen the last of oil and coal.[17] The world that is taking shape before our eyes nevertheless gives us reason to hope. More modest energy consumption will naturally stave off global tensions around the ownership of fossil-fuel sources, create green jobs in leading industrial sectors, and make Western countries serious energy contenders once again.[18] Irrespective of what the likes of Donald Trump, Jair Bolsonaro, and strategists in the Kremlin think, this transition is unstoppable: it involves big money that is pulling in players from across the economy — including the oil giants.

This energy transition traces its beginnings to Germany in the 1980s,[19] and culminated in Paris in 2015, when 195 nations jointly agreed to accelerate this formidable journey. Their goal: to keep the increase in global warming to below two degrees by the end of this century, mainly by replacing fossil fuels with green energy.

But just as the delegates were about to sign the Paris Agreement, a wise old man with pale-blue eyes and a bushy beard, dressed like a hermit descending from the mountain, entered the vast hall of COP 21. With an enigmatic smile on his face, he parted the amassed heads of state. Reaching the podium, he began to speak in a deep and deliberate voice: 'Your intentions are charming, and we can all rejoice in the new world to which you are about to give birth. But you are blind to the perils inherent in your audacity!'

Silence.

Turning to the Western delegates, he continued: 'This transition will cripple entire swathes of your economies, and the most strategic at that. It will plunge hordes of workers into retrenchment, triggering social upheaval that will shake your democratic foundations. Even your military sovereignty will be compromised.'

Now addressing all delegates, he added: 'The energy and

digital transition will devastate the environment in untold ways. Ultimately, the environmental price of building this new civilisation is so staggering that there is no guarantee you will succeed.'

He ended with an oracular message: 'Your power has blinded you to the point that you have lost the humility of the sailor before the ocean, the climber before the mountain. You forget that the Earth will always have the final say!'

The wise old man is, of course, a figment of my imagination. But the message is real enough, and crystal clear: the 196 delegations in Le Bourget that day signed the Paris agreement and committed to this Herculean task without considering a few crucial questions.[20] Where and how are we going to procure the rare metals without which this treaty will fail? Will there be winners and losers on the new chessboard of rare metals, as there were for coal and oil? And what will be the economic, social, and environmental cost of securing their supply?[21]

For eight years and across a dozen countries, I researched these new rare substances that are already upending our world. I ventured deep into mines in the tropics of Asia, eavesdropped on deputies in the corridors of the French National Assembly, flew over the deserts of California in a light aircraft, bowed before the queen of a tribal community in southern Africa, travelled to the 'cancer villages' of Inner Mongolia, and blew the dust off old parchments in venerable London institutions.

Across four continents, men and women involved in the opaque and underground world of rare metals shared with me a very different and far darker tale of the energy and digital transition. By their account, the emergence of these new substances in the wake of fossil fuels has not done us or the planet any of the favours we would expect from a supposedly greener, friendlier, and more insightful world — far from it.

Great Britain dominated the nineteenth century, thanks to its hegemony over global coal production. Many of the events of the twentieth century can be seen through the lens of US and Saudi Arabian control over oil production and supply routes. In the twenty-first century, one state is in the process of dominating the export and consumption of rare metals. That state is China.

Consider this economic and industrial observation: by committing to the energy transition, we have flung ourselves headlong into the jaws of the Chinese dragon. Arguably, the Middle Kingdom holds a near monopoly over a profusion of rare metals without which low-carbon and digital energies — the very foundations of the energy transition — cannot exist. And, as I will address later in this book, China has used barely credible chicanery to position itself as the sole supplier of the most strategic of the rare metals. Known as 'rare earths', they are difficult to substitute, and the vast majority of industrial groups cannot do without them.[22] (See Appendix 14 for the main industrial applications of rare earths.)

And so the West has placed the fate of its green and digital technologies — the cream of its industries of the future — in the hands of just one nation, while China is nurturing its own technologies and playing hardball with the rest of the world by putting a cap on the export of those resources. The result: serious economic and social consequences for the rest of the world.

Next, an ecological observation: our quest for a more ecological growth model has resulted in intensified mining of the Earth's crust to extract the core ingredient — rare metals — with an environmental impact that could prove far more severe than that of oil extraction. Changing our energy model already means doubling rare metal production approximately every fifteen years. At this rate, over the next thirty years we will need to mine more mineral ores than humans have extracted over the last 70,000 years. But the shortages already looming on the horizon could burst the bubble of Jeremy Rifkin, green-tech industrialists, and Pope Francis, and prove our hermit right.

The third observation relates to geopolitics and the military. The continued existence of the most sophisticated Western military equipment (robots, cyberweapons, and fighter planes, including the US's supreme F-35 stealth jet) also partly depends on China's goodwill. This has United States intelligence leaders concerned, as the US administration prepares for a potential war with China in the South China Sea.[23]

This latest scramble for resources is already heightening tensions over ownership of the most abundant deposits, sparking territorial conflicts in peaceful backwaters apparently of interest to no one. Fuelling this thirst for rare metals are a burgeoning global population set to reach 8.5 billion by 2030, the boom in new modes of high-tech consumption, and the growing convergence of Western and emerging economies.[24]

By seeking to break free from fossil fuels and turn an old order into a new world, we are in fact setting ourselves up for a new and more potent dependence. Robotics, artificial intelligence, digital healthcare, cybersecurity, medical biotechnology, connected objects, nanoelectronics, driverless cars ... the most strategic sectors of the economies of the future, all the technologies that will exponentially increase our computing capacity and modernise how we consume energy, our daily routines, and even our most significant collective choices will depend entirely on rare metals. These resources will provide the fundamental building blocks of the twenty-first century. Yet our addiction is already pointing to a future so far unpredicted. We thought we could free ourselves from the shortages, tensions, and crises created by our appetite for oil and coal. Instead, we are replacing these with an era of new and unprecedented shortages, tensions, and crises.

From tea to black oil, nutmeg to tulips, saltpetre to coal, commodities have been a backdrop to every major exploration, empire, and war, often altering the course of history.[25] Today, rare metals are changing the world. Not only are they polluting the environment, but they are jeopardising economic stability and global security. In the twenty-first century alone, their use has consolidated China's supremacy and accelerated the weakening of the West that began at the turn of the millennium.

But the rare metals war is far from lost. China has made some colossal errors; the West can respond; and the technical progress we have yet to make is bound to transform how we generate wealth and energy.

Until then, this book recounts the dark side of the story of the world that awaits us. It is an undercover tale of a technological odyssey that has promised so much, and a look behind the scenes of our lavish and ambitious quest that involves risks as formidable as those it sets out to resolve.

CHAPTER ONE

The rare metals curse

'WHAT DO YOU WANT? YOU HAVE NO BUSINESS HERE!' A MAN in his forties pulls up to us in a black Audi, and glares at us menacingly. He is joined by a companion, equally menacing, and soon another man on a motorbike also pulls up. 'You need to leave, it's dangerous. We don't want any trouble!' The three men start to lose their cool; the tension is palpable. 'Get lost!' shouts the man in the Audi. He can tell that I am trying to buy time by my furtive glances at the tent erected incongruously on the hillside.

'People still work here,' whispers Wang Jing, a former miner and my scout. 'I was sure they'd closed these quarries ages ago!' The scatter of new equipment and evacuation pipes in the area confirm my doubts. Two hundred metres away, the telltale tent stands above the tailing ponds and disembowelled rocky landscape. There can be no doubt that rare metal-refining activities are taking place at this camp. Where are they extracting the minerals from? 'From the mines all around us,

but also from the enormous illegal quarries flanking the hill,'
Wang Jing explains.

Two days earlier that July in 2016, I had landed at the tiny
airport of Ganzhou, a town in the Chinese province of Jiangxi
some 700 kilometres south of Beijing. From there, I drove due
south for hours on battered roads, hemmed in by row upon
row of rice paddies, to reach the mines. The last few dozen
kilometres took me along little more than ribbons of asphalt,
weaving between rickshaws and trailers laden with rubble, and
women wearing the traditional *co mào* conical hat. Out of the
foothills of the Nan Kang mountains burst lotus forests and
palm trees — a lush and abundant organic kingdom of stifling
foliage pushing into the blue sky.

I was also in the biggest rare metals mining area on the
planet.

Rare metals: a definition

When it comes to raw materials, nature can be surprisingly
generous or deeply parsimonious. Alongside popular species such
as the poplar and pine are rare trees like the hairy quandong in
Australia or the ghost orchid in the UK. Tulips may very well
overflow the fields of Holland, but other flowers, such as the
butterfly orchid, barely make an appearance in flower shops in
the Netherlands. In these parts, the skies abound with birds like
the mallard — much to the delight of hunters across Western
Europe. Then there are the more discreet, rarely sighted birds
like the California condor in North America.

Similarly, abundant metals like iron, copper, zinc, aluminium,
and lead coexist with a family of some thirty rare metals.' The

lists published by the United States Geological Survey (USGS), an agency of the United States Department of the Interior, and by the European Commission are an education in themselves: light and heavy rare-earth elements, germanium, tungsten, antimony, niobium, beryllium, gallium, cobalt, vanadium, tantalum, and other rare metals.[2]

They share the following traits:

- They are associated with abundant metals found in the Earth's crust, but in minute proportions. To give a clear idea, the most abundant elements in the Earth's crust are oxygen (46 per cent), silicon (28 per cent), aluminium (8 per cent), and iron (5.5 per cent). Neodymium and gallium are, respectively, 1,400 and 3,000 times less abundant than iron.[3]

- Naturally, this is reflected in the markets. Rare metal production is on a minute scale, and is overlooked by mainstream media: in 2022, around 300,000 tonnes of rare earths were produced, compared with 2.6 billion tonnes of iron — 9,000 times less. The same goes for gallium, of which 500 tonnes are produced worldwide, compared with 22 million tonnes of copper — 40,000 times less.[4]

- This makes these rare metals expensive: in 2022, 1 kilogram of gallium was worth around US$350, or nearly 3,000 times more expensive than iron. Germanium costs five to ten times more than gallium![5]

- They also possess the exceptional properties demanded by new — and especially 'green' — technology manufacturers working to reduce our carbon footprint on the environment.[6]

Rare metals: drivers of new energies

Since the dawn of time, humans have sought to transform sources of natural energy (such as wind, thermal, and solar) into mechanical energy. Take the windmill, for instance. Its vanes and rotor are driven by wind energy to actuate a mechanical mill that then crushes olives or grain. In the steam engine, thermal energy transported by steam from water is converted, using pistons, into mechanical energy powerful enough to drive a locomotive. Thermal energy is also generated by the combustion of fuel to drive the pistons of a vehicle and set it in motion. In essence, we have been making movement-generating machines for centuries.[7] The more we multiply the possibilities of movement, the more we can travel and trade, entrust new tasks to machines and robots, and make productivity gains — and therefore greater profits.

Energy needs to be both abundant and inexpensive to ensure that machines run efficiently — a challenge we must overcome to satisfy our economic-growth ambitions. Thus, for almost three centuries we have been working tirelessly at developing new engines with increasingly impressive power-to-weight ratios: the more compact and less resource-intensive they are, the greater their mechanical energy output.

Enter rare metals. While mineralogists have known of their existence since the eighteenth century, they garnered little interest while their industrial applications remained undiscovered. But from the 1970s, humans began to exploit the exceptional magnetic properties of some of these metals to make super magnets.[8]

An electrical charge flowing through the stator coils of an electric motor creates a rotating magnetic field that drives the permanent magnets in the rotor. The smallest of these magnets is barely the size of a pinhead, while the biggest magnet ever designed has an external diameter of five metres, weighs 132 tonnes, and is located at the Saclay Nuclear Research Centre in the Paris region.[9] Irrespective of their size, magnets are to electric motors what pistons are to steam and petrol engines. Magnets have made it possible to manufacture billions of engines, both big and small, capable of executing certain repetitive movements in our stead — whether it be running a motorbike, powering a train, making an electric toothbrush or mobile phone vibrate, operating an electric window, or launching an elevator to the top of the tallest skyscraper. It is estimated that the world will need ten to twelve times as many magnets by 2030 for electric vehicles alone.[10]

Without realising it, our societies have become completely magnetised. To say that the world would be significantly slower without magnets containing rare metals is not an understatement.[11] This is something to think about when rearranging your holiday magnets on the fridge!

The technological revolution behind the energy shift

Electric engines did more than make humanity infinitely more prosperous; they made the energy transition a plausible hypothesis. Thanks to them, we have discovered our ability to maximise movement — and therefore wealth — without the use of coal and oil. It is not surprising that electric engines will soon replace conventional engines. Electric engines are already being used to propel ships, send the Solar Impulse aircraft around the world, launch space probes and satellites, and put enough electric cars on the roads to disrupt the automotive market.[12]

Of course, these engines are connected to electric batteries that create the electricity needed to power the stator coils whose magnetic field drives activate the rotor's magnets. The difference, however, is that with rare metals it is possible to generate clean energy: they cause the rotors of certain wind turbines[13] to turn and convert the sun's rays into electricity using solar panels.[14] Because they remove pollution from most of the energy cycle — from manufacture to end use — we can safely envisage a world without nuclear, oil-fired, or coal-fired power plants.

But that is merely scratching the surface of rare metals, for they possess a wealth of other chemical, catalytic, and optical properties that make them indispensable to myriad green technologies.[15] An entire book could be written on the details of their characteristics alone. They make it possible to trap car-exhaust fumes in catalytic converters, ignite energy-efficient light bulbs,[16] and design new, lighter, and hardier industrial equipment, improving the energy efficiency of cars and planes. Two thousand years ago, the Hebrews were able to cross the Sinai desert surviving on manna, the providential food sent from heaven. Today, another godsend — this time from underground — has been laid out at the ecological banquet, for there is a rare metal for every green application. Surely there is a green guardian angel watching over us.

Most surprising is how these metals have become indispensable to new information and communication technologies for their semiconducting properties that regulate the flow of electricity in digital devices. And the once-distinct functions of green and digital technologies are beginning to converge. Indeed, increasingly sophisticated software and algorithms used in 'smart grids' make it possible to regulate fluctuations in the flow of electricity between producers and consumers. This is precisely what the 80 million smart meters already installed in the US are doing. In the smart cities of tomorrow, which will combine green and digital technologies, we will save up to 65 per cent of the electricity we use today, thanks to sensor-embedded streets that adjust the lighting to foot traffic, while weather-prediction software makes solar panels 30 per cent more efficient.

Thus, digitalisation and the energy transition are co-dependent. Digital technology advances and enhances the impacts of green tech. Their combination is ushering in an era of energy abundance, stimulating new industries, and has already created over 12 million jobs worldwide.[17] This is a boon not lost on political leaders: to help these new markets take off, Europe is now urging its member states to reduce their carbon dioxide emissions by 55 per cent (in relation to 1990 levels) by 2030, and to increase to 32 per cent the renewable energy share of their energy consumption.[18]

But why stop there? According to négaWatt, a French non-governmental think tank, renewable sources of energy could meet all of France's energy needs by 2050.[19]

The acceleration of rare metal consumption

This technological diversification has multiplied the types of metals that humanity uses. Between the ages of antiquity and the Renaissance, human beings consumed no more than seven metals;[20] this increased to a dozen metals over the twentieth century; to twenty from the 1970s onwards; and then to almost all eighty-six metals on Mendeleev's periodic table of elements. (See Appendix 1.)

Our appetite for metals boomed — and it didn't stop there. On the one hand, consumption of the three main sources of energy currently used in the world (coal, oil, and gas) tends to stabilise, decrease, or, at best, moderately increase.[21] On the other hand, the potential demand for rare metals is exponential. We are already consuming over two billion tonnes of metals every year — the equivalent of more than 500 Eiffel Towers a day.[22] (See Appendix 2 to see the trends in world primary metal production, and more specifically Appendix 3 on the production of rare earths.) However, forecasts by the International Energy Agency indicate that by 2040, demand for rare earths could be multiplied by seven, nickel by nineteen, cobalt by twenty-one, graphite by twenty-five, and lithium by forty-two, compared with 2020 needs.[23] (See Appendix 4 for the projected growth in rare metal and mineral needs in 2040 versus 2020.) By 2035, demand is expected to double for germanium; quadruple for tantalum; and quintuple for palladium. The scandium market could increase nine-fold, and the cobalt market by a factor of 24.[24] There is going to be a scramble for these resources, for the resilience of capitalism relies increasingly on the emergence of green and digital technologies. The market will become less and less dependent on the fuels of the last two industrial revolutions, and will increasingly rely on the metals that are driving the impending transition.

The US Geological Survey and the European Commission agency in charge of raw materials have produced a map of the world's rare metal production areas. It shows that South Africa is a major producer of iridium, platinum, and rhodium; Russia, of palladium; the US, of beryllium; Brazil, of niobium; Turkey, of borates; and the Democratic Republic of the Congo (DRC), of cobalt, tantalum, etc. Yet most of these metals come from Chinese mines. This is the case for germanium, bismuth, magnesium, and, above all, the supreme 'green' metals, whose staggering electromagnetic, optical, catalytic, and chemical properties surpass all others in performance and fame: the rare-earth metals.

They form a family of seventeen elements, featuring exotic names like scandium, yttrium, lanthanum, cerium, praseodymium, neodymium, samarium, europium, gadolinium, terbium, dysprosium, holmium, erbium, thulium, ytterbium, lutetium, and promethium. (See Appendix 5 for a map of rare-mineral-producing countries.)

Rare earths, the black market, and environmental disasters

The biggest quantity of rare earths is extracted from the bowels of Jiangxi, in the heart of tropical China, which is where our story begins.[25]

Wang Jing knows this better than anyone. I had met the fresh-faced 24-year-old, with smiling eyes under his mop of hair, in the village of Xing Quang. He knows this strip of mountain like the back of his hand, and had little difficulty guiding me there. In fact, he spent years working at this illegal mine concealed by a copse of eucalyptus trees. He tells me how he would chip away at red-tinted rock and crush prodigious rubble aggregates alongside other miners, both men and women.

Like a human anthill, the mountain was mined twenty-four hours a day, seven days a week. Miners were paid a few hundred euros a month, and slept on ground plundered by picks and excavators. At this frenetic pace, hundreds of thousands of tonnes of minerals were extracted from the mountain. But four years earlier, the Chinese authorities had banned these activities, and illegal miners were slapped with heavy fines. Entire stocks of metals promised to foreign markets were seized at the port of Canton, some hundreds of kilometres south, and dozens of traffickers were thrown into prison.

Despite this, the more determined and needy miners have entrenched themselves in the folds of the mountain's most inhospitable terrains. They prosper in secret, and are said to make payoffs to the local police. Their activities are feeding a colossal Chinese black market for minerals that, once processed, are exported worldwide.

These are the activities I have come to see. And the three illegal miners standing in our way know it. The motorcyclist threatens me again. I move away from the tent; clearly, I won't get to see what I came for: proof of the staggering pollution created by rare-earth mining.

'It's poison,' says Wang Jing. 'The chemicals used for refining the minerals were poured straight into the ground.' The sulphuric and hydrochloric acids would pollute the nearby stream to the point that 'it was impossible for any plant to grow'. Because the closest housing lies far from the Yaxi mountains, there was no visible impact on residents. But elsewhere, housing was much closer, he said.

The 10,000 or so mines spread across China form roughly the size of Lebanon, and have played a big role in destroying the country's environment.[26] Pollution damage by the coal-mining industry is well documented. But barely reported is the fact that mining rare metals also produces pollution, and to such an extent that China has stopped counting contamination events. In 2006, some sixty companies producing indium — a rare metal used in the manufacture of certain solar-panel technologies — released tonnes of chemicals into the Xiang River in Hunan, jeopardising the meridional province's drinking water and the health of its residents.[27] In 2011, journalists reported on the damage to the ecosystems of the Ting River in the seaside province of Fujian, due to the operation of a mine rich in gallium — an up-and-coming metal for the manufacture of energy-efficient light bulbs.[28] And in Ganzhou, where I landed, the local press recently reported that the toxic waste dumps created by a mining company producing tungsten — a critical metal for wind-turbine blades — had obstructed and polluted many tributaries of the Yangtze River.

A Chinese journalist reporting anonymously describes the working conditions — reminiscent of a bygone era — at the graphite mines of Shandong, in eastern China. In the processing plants rising out of dark, uprooted mounds of the Earth's crust, '[M]en and women, wearing no more than basic face masks, work in areas thick with black particles and acid fumes. It's hell.' To complete the picture are toxic pits of chemical discharges from the plants, fields of poisoned corn, acid rain, and more. 'Local authorities tried to police environmental offences,' says the journalist, 'but the pressure from automobile manufacturers was too great.'

Dirty metals for a greener world

The assertion that producing the metals we need for a cleaner world is in fact a polluting process seems incomprehensible at first. Which is understandable: most consumers have forgotten what they learned in their high-school natural sciences, physics, and chemistry classes. Let's refresh our memories.

No need to dust off the chalkboard; a trip to the closest bakery will do. Everyone knows the ingredients of a loaf of bread: a good portion of flour, water, a bit of yeast, and a pinch of salt. It's not that different from a rock of a similar size taken out of a mine: its ingredients comprise several minerals mixed together.

In this metaphor, the flour represents the rock that ends up on the rubble heap. The water is where it starts to get interesting: all things being equal, it represents iron — a mineral found in abundance in the Earth's crust. Next is the yeast, making up a much smaller portion of the mix, which represents nickel — a metalloid rarer than iron. That leaves the pinch of salt: our rare metals. Their concentration in the Earth's crust is as minute and imperceptible as the pinch of salt sprinkled into our bread dough.

But rock is composed of minerals aggregated over billions of years, and the rare metals are therefore completely incorporated in the rock — just like the salt when it is kneaded and baked into the dough. You would think that trying to extract it would be practically impossible, yet decades of research have developed the chemical processes to do just that. And the Chinese 'sorcerer's apprentices' deep in the mines of Jiangxi province and elsewhere are managing to achieve this: to extract rare metals from rock.

For a process known as 'refining', there is nothing refined about it. It involves crushing rock, and then using a concoction of chemical reagents such as sulphuric and nitric acid. 'It's a long and highly repetitive process,' explains a French specialist. 'It takes loads of different procedures to obtain a rare-earth concentrate close to 100 per cent purity.'[29]

That's not all: purifying a single tonne of rare earths requires using over 30,000 cubic metres of water, which then becomes saturated with acids and heavy metals.[30] Will this water go through a water-treatment plant before it is released into rivers, soils, and groundwater? Very rarely. The Chinese could have opted for clean mining, but chose not to. From one end of the rare metals production line to the other, virtually nothing in China is done according to the most basic ecological and health standards. So as rare metals have become ubiquitous in green and digital technologies, the exceedingly toxic sludge they produce has been contaminating water, soil, the atmosphere, and the flames of blast furnaces — representing the four elements essential to life. The result is that producing rare metals has become one of the most polluting — and secretive — industries in China. But that won't stop me from taking a closer look.

My next stop was Hanjiang, a few dozen kilometres from the rare-earth mines I surveyed with Wang Jing. It is a hamlet located close to another of these mines. But 90 per cent of its inhabitants had fled the jumble of stone houses and their dark-tiled roofs. The residents complained that because of the rampant mining activities, '[n]othing we plant grows anymore. Our rice paddies have become infertile!' Those who have refused to leave have accepted their fate. 'What can we do?' asked an old man, overwhelmed by the thick, cloying air. 'There's no point even complaining about it.' Do the local authorities know about the pollution? 'Of course they do! Even you would have guessed without anyone telling you!'

The heavy toll on health

This is nothing compared to what awaits me 2,000 kilometres north in Baotou, the capital of the autonomous region of Inner Mongolia, which I first visited in 2011. It's a city well known to all rare metal hunters for the simple reason that it is the biggest rare-earth production site on the planet, far surpassing Jiangxi province. There I saw convoys of trailer trucks laden with gravel trundle down the dusty roads of the city and surrounding countryside. The hundreds of thousands of tonnes of rare earths extracted annually at the Bayan Obo site by the mining giant Baogang — responsible for 45 per cent of global production — contributes to the prosperity of the city and its three million or so inhabitants.[31]

It must be said that I find Baotou quite pleasant, with its flurry of Chinese flags waving from the roofs of buildings, and swarms of bicycles zipping between the city and its industrial areas, while the troubled waters of Asia's second-longest river, the Yellow River, caress the city's edges. At the entrances of the city's parks are hundreds of posters depicting a couple and their child against a green, pristine background bearing the slogan: *Building a clean city for our country*. It's postcard perfect.

It is impossible, however, to get anywhere near the Baogang mines, some 100 kilometres from the city centre. Having already been marched to the police station by a pair of overzealous police officers, I was in no hurry to go back. But my Chinese fixer reckoned that by going just a few dozen kilometres west of the city, I could catch a glimpse of the industry's secretive activities.

Past the suburbs of Baotou, below a quadruple carriageway, a lonely path led me to a cement embankment bristling with pylons, each one equipped with a security camera to watch for intruders. This is how I reached the Weikuang Dam — an artificial lake into which metallic intestines regurgitate torrents of black water from the nearby refineries. I was looking at 10 square kilometres of toxic effluent, which occasionally flows over into the Yellow River.[32]

This is also the beating heart of the energy and digital transition.

I was left speechless during the hour I spent observing this immense, disintegrating lunar landscape. Wang Jing and I decided to get moving before the security cameras alerted the police to our presence.

A few minutes later, we arrived in Dalahai, on another side of the artificial lake. In this village of redbrick houses, where the thorium concentration in the soil is in some places thirty-six times higher than in Baotou, the thousands of villagers still living there breathe, drink, and eat the toxic discharge of the tailings dam.[33] We met one of them, Li Xinxia. With her striking features and wistful eyes, the 54-year-old woman knew this was a touchy subject, but confided in me anyway: 'There are a lot of sick people here. Cancer, strokes, high blood pressure ... almost all of us are affected. We are in a grave situation. They did some tests here, and our village was nicknamed "the cancer village". We know the air we are breathing is toxic and that we don't have that much longer to live.'

Is there any way out for Li Xinxia and her loved ones? The provincial authorities did offer the villagers 60,000 yuan per mu of land (around US$9,000 for 666 square metres) to relocate to high-rises built in a neighbouring town. While it was a handsome sum in a rural area where the average annual income is around US$1,300, it was not enough to enable the farmers to make ends meet. The apartments were prohibitively priced for people who could no longer live off land that had become infertile.

Rare earths have cost the community dearly. The hair of young men barely thirty years of age has suddenly turned white. Children grow up without developing any teeth. In 2010, the Chinese press reported that sixty-six Dalahai residents had died of cancer.

I return to Baotou in spring 2019. The town has expanded considerably, and its suburbs now encroach on a succession of huge rare-earth refineries. According to a few people we questioned discreetly on the site's perimeter, the industrial group had destroyed everything before it extended its operations — starting with Dalahai, where the 'cancer villages' were razed and their inhabitants compensated to move. All that remains are pieces of bricks that men and women from neighbouring villages come and clear at the end of the afternoon when the heat is less oppressive.

Weikuang Dam still lies between large man-made embankments, and is fed incessantly by factory effluents. 'The reservoir is massive. If you look at it from the roof of a building, you'll understand. It's so big you'd think it was the sea!' The presence of Chinese security nearby does not stop Gao Xia from giving us her testimony. The 48-year-old villager has been rehoused with her husband in a gloomy high-rise estate overlooking the devastated landscape. Eight years after I first covered the story, the same causes seem to be producing the same effects. By Gao Xia's account, it is a disaster area. The water in the rivers 'is whitish-green and sometimes red', and the land produces corn and buckwheat with great difficulty, while cancer continues to affect local populations. After a life of living off the land, Gao Xia is condemned to eke out a living doing odd jobs here and there. She speaks of her 'bitterness' at being powerless against the rare-earth companies 'that have polluted our environment'.[34]

'The Chinese people have sacrificed their environment to supply the entire planet with rare earths,' Vivian Wu, a recognised Chinese expert in rare metals, tells us. 'Ultimately, the price of developing our industry is just too high.'

How could Beijing have allowed this disaster to happen?

Playing catch-up at the risk of anarchy

To answer this question, we need to go back in time. The nineteenth and twentieth centuries were times of decline and humiliation for the Middle Kingdom. At the death of the Qianlong Emperor – the 'Chinese Louis XIV' – in 1799, China was the global powerhouse. The empire's borders reached the furthermost bounds of Mongolia, Tibet, and Burma. More clement temperatures and bountiful harvests led to a population boom, and at the zenith of the Qing dynasty, the political system was stable and the country's economic production represented one-third of global gross domestic product (GDP). Middle Kingdom mania extended as far as Europe: the French writer, historian, and philosopher Voltaire waxed lyrical on the merits of Manchurian autocracy, *chinoiserie* was all the rage, and the English discovered their love of tea.

But soon the edifice crumbled, followed by one disaster after another: opium wars,[35] unfair treaties, humiliation at the Treaty of Versailles in 1919 (despite China being one of the victors of the First World War),[36] the failures of the Kuomintang party,[37] and the devastating effects of Maoism. At the death of Mao Zedong in 1976, China's position in the world economy had diminished tenfold compared to its position at the end of the eighteenth century. The country had been ravaged by civil wars, and the survivors of the bloody Cultural Revolution (which killed millions) were subjected to ghastly brainwashing.[38]

But the Chinese are resilient; their hunger to recover their lost prestige is insatiable. After all, between the year 960 and today, China was the leading global power for close to nine centuries. The Middle Kingdom had to take back its place — at all costs.

Obsessed by the idea of erasing the failings of the nineteenth and twentieth centuries as quickly as possible, China has been racing at reckless speed to achieve in three decades the economic progress that took the West three centuries to accomplish. In 1976, the Communist Party, under the leadership of Deng Xiaoping, opened the country to capitalism and global trade. Its policy of combining economic and environmental dumping in the form of below-market-value prices and lax environmental rules gave China a competitive advantage over Western countries, making it the factory of the world and the West's official supplier of low-cost goods. Lastly, and most importantly, Beijing became the primary producer of all the minerals that the world needs to support its economic growth. Today, China is the leading producer of thirty-three of the fifty-one mineral resources that are vital to our economies, often representing over 50 per cent of global production.[39]

The downside of this spectacular success? Little attention has been given to the environmental impact of these economic choices. Industry has been left to pollute the atmospheres of major cities, to contaminate the soil with heavy metals, and to dump its mining waste in most rivers with impunity. Under the growth measures in place, anything goes. In other words, the Chinese have made a real mess of it.

The environmental cost is exorbitant, inhumane, and outrageous.[40] China is the biggest emitter of greenhouse gases (producing one-third of global greenhouse-gas emissions in 2022), and the alarming figures coming out of the country are multiplying.[41] Over 10 per cent of its arable land is contaminated by heavy metals,[42] and 80 per cent of its groundwater is unfit for consumption.[43] Only five of the 500 biggest cities in China meet international standards for air quality, and there are 2 million deaths per year due to air pollution alone.[44] In the words of Chinese environmental activist Ma Jun, whom I met in Beijing: 'This was a monumental error.'

The scourge of rare metals gone global

The pollution caused by rare metals is not limited to China. It concerns all producing countries, such as the Democratic Republic of the Congo, which in 2022 supplied nearly 70 per cent of the planet's cobalt, or 130,000 tonnes annually.[45] This resource — indispensable to the lithium-ion batteries used in electric vehicles — is mined under conditions straight out of the Middle Ages. Up to 300,000 miners equipped with spades and picks dig into the earth to find the mineral, especially in the southern region of Lualaba.[46] Given the DRC government's inability to regulate the country's mining activities, the pollution of surrounding rivers and turmoil in the ecosystems are legion. Research by Congolese doctors has found that the cobalt concentration in the urine of the local communities living near the mines of Lubumbashi, in Katanga province, and Kolwezi, the main city in the neighbouring province of Lualaba, is up to eleven times higher than in a control sample. In Kolwezi, samples taken from the surface of the ground contain on average of seventy times more cobalt than the control area studied, and thirteen times more uranium.[47]

We see the same in Kazakhstan, a central Asian country that produces 14 per cent of the world's chrome — prized by the aerospace industry for the manufacture of superalloys that improve the energy performance of aircraft.[48] In 2015, researchers from South Kazakhstan State University discovered that chrome mining was responsible for the colossal pollution of the Syr Darya, the longest river in Central Asia. Its water had become completely unfit for consumption by the hundreds of thousands of inhabitants, who are now even advised against using it for their crops.[49] Other studies have since been carried out to measure heavy metal concentrations in rivers. In 2020, a study by the Baishev University (Kazakhstan) and Samara University (Russia) showed that heavy metal levels were increasing year after year in the rivers studied, the Elek and the Or, both tributaries of the Ural River.[50]

Latin America has already started to experience similar

problems with lithium mining — a white metal lying below the salt flats of Bolivia, Chile, and Argentine. It is considered critical by the European Commission and the US government, and demand is expected to soar on the back of the electric car boom that has jacked up its global production. Naturally, Argentina has its sights set on becoming the giant of lithium. While there were only two mines in operation in in 2022, there are over thirteen mining projects in the pipeline, and dozens more under review. By 2025 the country could be producing up to 150,000 tonnes of lithium a year, or more than 50 per cent of global demand.[31]

In May 2017, all the rare metal exploration, mining, and refining companies operating in Latin America met on the banks of Rio Plata near Buenos Aires for the Arminera international mining trade fair. Amid the excavators, skips, light towers, and other waste-water treatment equipment on display, Daniel Meilán, Argentina's mining secretary, boasted about the 'dozens of prospecting activities for lithium deposits in progress' in the country, and promised a mining sector that would be responsible and compliant with international ecological standards. Before popping the champagne, and to the applause of those present, all the Argentinian industry players were invited to sign an ethics charter.

At the same time, some thirty Greenpeace activists had blocked the entrance to the trade fair, brandishing banners calling out the lies of the mining industry. 'Everything they say is pure greenwashing,' said one of its members, Gonzalo Strano. 'There's no such thing as sustainable mining. Not only does it dig out the ground by definition, it uses chemicals and massive amounts of water, which is a problem.'

The mining sector in Latin America has a sulphurous reputation. From Mexico to Chile, from Colombia to Peru, the last few years have seen growing opposition from local communities.[52] Most emblematic of this deep distrust is the Pascua Lama gold and silver mine, operated by Canadian mining group Barrick Gold, in the north of Santiago, Chile. Extracting at this site would have involved destroying the glacier concealing the orebody — a prospect that had residents up in arms, forcing Barrick Gold to shut down its activities in 2013.[53]

The Pascua Lama example inspired the entire Latin American mining sector. Large-scale lithium mining now sparks environmental activism. As with any mining activity, it requires staggering volumes of water, diminishing the resources available to local communities living on water-scarce salt flats. The communities of the Hombre Muerto salt flat in Argentina claim that pumping water has dried out the meadows used to graze their cattle. Along the Patos river, which springs from a small lake to the south-east of the salt flat, communities are also campaigning to protect their wells.[54]

Mining is not only dangerous for the environment, but for mine workers too. In 2019, at the Sudbury nickel mine in Ontario (nickel is an essential metal in the manufacture of batteries for electric vehicles), thirty-nine miners were trapped for forty-eight hours after a load obstructed the mine elevator.[55] The inherent risks of the job prompted miners to demand better working conditions. But confrontation can sometimes lead to tragedy. Such was the case at the Marikana platinum mine in South Africa. (Platinum is an essential component of hydrogen fuel cells.) In 2012, a strike lasted several weeks before the police intervened. Thirty-four miners were killed. The families of the victims are still trying to understand the actions of the mining company and the police forces. They feel they have received very little in the way of compensation, and are left with a lingering sense of injustice.[56]

When not creating injustice, extracting minerals from the ground is an inherently dirty operation. The way it has been carried out so irresponsibly and unethically in the most active mining countries casts doubt on the virtuous vision of the energy and digital transition. Given China's dominant role in the global supply of rare metals, we cannot accurately assess the progress made in combating global warming without properly accounting for Beijing's ecological performance. Which is catastrophic, to say the least.

The message of this overview of the environmental impacts of extracting rare metals from the Earth is clear: we need to be far more sceptical about how green technologies are manufactured. Before they are even brought into service, the solar panel, wind turbine, electric car, or energy-efficient light bulb bear the 'original sin' of its deplorable energy and environmental footprint. We should be measuring the ecological cost of the entire lifecycle of green technologies — a cost that has been precisely calculated.

CHAPTER TWO

The dark side of green and digital technologies

THE TECHNOLOGIES THAT WE DELIGHT IN CALLING 'GREEN' may not be as green as we think. They could even have an enormously negative impact on the environment, as we learned in Toronto in the spring of 2016.

In the heart of the city's financial district, everyone who's anyone in the North American mining industry — mining companies, experts, public authorities, venture capitalists, consultancies, and academics — gathered in the lavish setting of a grand hotel for a conference on the rare metals 'gold rush'.[1] On the agenda: investments, cashflow, gross margins, capital raising, cost structuring, market capitalisation, and average annual output. With the International Energy Agency expecting the share of renewables in global electricity output to increase from 28 per cent to 61 per cent by 2040, the green-tech growth outlook looks bright indeed.[2]

But in the midst of this predominantly male and well-heeled

crowd were two people about to throw a spanner in the mining works.

Green tech's heavy toll on the environment

The first was Canadian Bernard Tourillon, director of Uragold — a company that makes equipment for the solar industry. Tourillon has painstakingly calculated the ecological impact of photovoltaic panels: on account of their silicon content, producing just one panel generates as much as 70 kilograms of carbon dioxide. The 23 per cent annual increase in the number of solar panels over the next few years will increase their power-generation capacity by 10 gigawatts every year — but will also generate 2.7 billion tonnes of carbon emissions into the atmosphere, or as much pollution as over 1 billion vehicles.[3]

Thermal solar panels make an even bigger splash, with some consuming as much as 3,500 litres of water per megawatt hour.[4] That's three-and-a-half times more water than the requirements of a gas-fired power station.[5] More problematic still is the fact that solar farms are more often than not located in water-scarce areas.

The second party pooper was John Petersen, a Texan lawyer with an extensive career in the electric battery industry. Based on number crunching, countless academic papers, and his own research, Peterson reached an astonishing conclusion.

Rewind to 2012, when researchers at the University of California Los Angeles (UCLA) compared the carbon impact of a conventional fuel-driven car against that of an electric car.[6] Their first finding was that the production of the supposedly more energy-efficient electric car requires far more energy than the production of the conventional car. This is mostly on account of the electric car's very heavy lithium-ion battery. The battery of prominent US company Tesla's Model S, for instance, weighs in at 544 kilograms — half the weight of a Renault Clio.[7] (See Appendix 6 for an overview of the metals used in an electric vehicle, and Appendix 7 for the material intensity of an electric car compared with that of a conventional car.)

Then there's the composition of the lithium-ion battery: 20 per cent graphite; 10 per cent nickel; 8 per cent copper; 3 per cent cobalt, lithium, a small amount of manganese; and, of course, steel (15 per cent) and aluminium (10 per cent).[8] By now, we are familiar with the conditions in which these minerals are extracted in China, Kazakhstan, and the Democratic Republic of the Congo. But we also need to consider how these minerals are refined, not to mention the logistics of their transportation and assembly. The UCLA researchers reached the conclusion that the large-scale manufacture of electric vehicles is more energy-intensive than that of conventional cars.[9]

Looking at its end-to-end lifecycle, however, the electric vehicle has the undeniable advantage of not requiring petrol. This brings down its carbon dioxide emissions into the atmosphere, offsetting its initial footprint. The French Environment & Energy Management Agency (ADEME) estimates that the cumulative emissions of an electric vehicle are around 13 tonnes of carbon dioxide equivalent, compared with more than twice that of its conventional counterpart.[10] It does point out, however, that these low-carbon properties depend on the battery's capacity, which must be less than 60 kWh, giving a range of 450 kilometres. For an electric SUV with a 100-kWh battery, the cumulative emissions are 18 tonnes of carbon dioxide equivalent, which is still less than for a conventional vehicle, but which considerably reduces the environmental benefits of electric cars. And some of the Tesla Group's Model 3s now have batteries capable of delivering a range of 630 kilometres.[11]

We also need to consider the other sources of pollution generated by electric vehicles. A study published in 2018 by researchers at the University of Florence in Italy is particularly insightful in this respect, as it compares the impact of electric and combustion-powered cars based on more diverse ecological criteria such as the acidification of ecosystems, particle emissions, human toxicity, and the depletion of resources. The results show that the electric car has a greater negative impact than its combustion-powered counterpart in all four of these much-overlooked criteria.[12]

The conclusions drawn in 2016 by John Petersen remain relevant: 'electric vehicles may be technically feasible, but their manufacture will never be environmentally sustainable'.[13] In any scenario, studies show that it is now essential to rethink the use of vehicles by reducing their mass and extending their lifespan ... Something that nobody seems prepared to do today.

This leaves us with a list of unanswered questions. Is the replacement impact of EV batteries, which have a short lifespan, being taken into account? Do we know the precise ecological cost of all the electronics and connected components packed into these electric vehicles? What about the environmental impact of recycling these mostly still-new vehicles in the future? And how much energy will it take to build the necessary electric grids and plants to meet these new needs?

Ultimately, as was quietly admitted to me by a US rare metals expert in Toronto that day, 'No one in the green-energy business is going to communicate on these points. Besides, don't we all want to believe that we're making things better rather than making things worse?'

Tangible invisibility

It doesn't stop there. We know that green technologies are gradually converging with digital technologies, thanks to which, their champions proclaim, green technology will be ten times more effective. This raises the legitimate question of whether these technologies will in turn make green-tech pollution worse. This is certainly not what proponents of the energy transition extol. Digital technologies, they tell us, will in fact moderate our energy consumption. Such is their overriding message, which I will now carefully unpack:

- First, digital technologies form the basis of 'smart' electrical grids, which are designed to optimise our energy consumption. Between solar panels that generate clean energy and 'zero-emission' cars that use energy without polluting, there need to be grids to supply energy. With the standard grid, the electricity generated by coal-fired, oil-fired, and nuclear power plants feeds continuously into the grid. Because we determine output, we know almost exactly how much energy passes at any given time and at any point in the grid. There is nothing of the sort with the current energy transition, which draws on what is known as 'intermittent' energy sources — intermittent because no one has worked out how to control the sun and the wind. The power supplied to the grid by solar panels and wind turbines is sporadic. Grid operators are therefore faced with the task of directing the right amount of electricity to the right place at the right time. If there is not enough electricity, supply comes to a standstill. If there is too much electricity, the surplus is wasted.

Energy operators are therefore delighted by the prospect of a new generation of electric grids that can constantly and dynamically manage supply in response to actual demand — thereby reducing energy wastage — using increasingly sophisticated algorithms.

- Digital technology supposedly mitigates the carbon impact of human activities, as lauded by the invigoratingly optimistic writings of new-technology advocates such as Jeremy Rifkin.[14] He postulates that the crossover of digital technologies and green energy will allow anyone to produce their own clean, inexpensive, and abundant electricity. The techno-prophet returned to the scene a few years later with an incredible idea: the new 'zero marginal cost society'.[15] By creating a new generation of 'collaborative commons' whereby everything is shared and exchanged online, internet technologies will pitch us into the age of access trumping ownership. No longer will we need to own anything; by simply surfing the web, we will have the freedom to share any product in exchange for money. Already we are seeing the beginnings of this cultural revolution in car transport (Blablacar, Getaround, to name but a couple). The consequences could be grave for the automotive industry. According to Rifkin, 80 per cent of car-sharing website users have already sold their car. Just think of the plummeting numbers of cars in this new 'Age of Access', and the commodity and carbon-emission savings that will go with it!'[16]

- In 2013, Eric Schmidt, the then chair of Google's board of directors, and Jared Cohen, a former advisor to Hillary Clinton in the US State Department and the self-appointed father of 'digital diplomacy', took the rationale a step further with the publication of their book *The New Digital Age*.[17] The global bestseller has helped open our eyes to the growing role of the virtual world. Courtesy of the internet, the two gurus explain, 'the vast majority of us will increasingly find ourselves living, working, and being governed in two worlds at once': the physical world and the virtual world. In the future, there will be more and more cyber-states, declaring more cyber-wars against virtual criminal networks that perpetrate increasingly powerful cyber-attacks.[18]

Yet this is also a prophecy that promises the utopia of a dematerialised world. Already dematerialisation is synonymous with working from home, e-commerce, electronic documentation, digital data storage, and more. By limiting the physical transportation of information, and migrating from paper to digital, we can disavow our resource-guzzling civilisation

and, while we're at it, slow down the deforestation
of the Amazon and the Congo Basin.[19] In a nutshell,
we are parachuting into a wiser, more moderate age.

But digital technology requires vast quantities of metals.
Every year, the electronics industry consumes 6 per cent of the
global demand for gold (around 300 tonnes), and 20 per cent
of the global demand for silver (around 7,000 tonnes).[20] Digital
technologies also guzzle a large portion of the global production
of rare metals: 15 per cent of palladium, 40 per cent of tantalum,
41 per cent of antimony, 42 per cent of beryllium, 66 per cent of
ruthenium, 70 per cent of gallium, 87 per cent of germanium,
and as much as 88 per cent of terbium.[21] This excludes the other
forty or so metals, on average, contained in mobile phones. (See
Appendix 8 for the rare metals composition of a mobile phone.)
And yet 'the product in the hands of the consumer only makes
up 2 per cent of the total waste generated over the course of the
product's lifecycle', explain the authors of a French book that
delves into the dark side of digital technology.[22] One example
says it all: 'Manufacturing one 2-gram chip alone produces 32
kilograms of waste' — a 1:16,000 ratio between the end-product
and the resulting waste.[23]

And this is only the manufacture of digital devices.
Operating electric grids will, of course, generate additional
digital activity — and therefore additional pollution, the effects
of which are becoming clearer. A documentary investigating the
environmental impact of the internet traces the journey of a
simple email: once it leaves the computer, it reaches the modem,
travels down the building to a connection centre, transits from
an individual cable to national and international exchanges, and

then goes through a message host (usually in the US). In the data centres of Google, Microsoft, and Facebook, the email is then processed, stored, and sent to its recipient. All told, our email travels 15,000 kilometres at the speed of light.[24]

All this has an environmental cost. In 2011, ADEME put an estimate on the energy cost of our digital activities: an email with an attachment uses around 100 Wh, or the equivalent of a low-energy light bulb left on for 12 hours.[25] Not all emails are that big, but it's staggering, considering that over 300 billion emails are sent every day.[26] One data centre alone uses as much energy as a city of 30,000 inhabitants to manage the flow of data and run its cooling systems.[27]

More broadly, a recent study by the French non-profit Green IT [IT for information technology] and the NegaOctet repository estimates that all digital services in Europe consume almost 10 per cent of Europe's electricity, and emit 185 million tonnes of carbon dioxide equivalent, compared with around 130 million tonnes for air travel.[28] This is just the tip of the iceberg, for the energy and digital transition will require constellations of satellites — already promised by the heavyweights of Silicon Valley — to put the entire planet online. It will take rockets to launch these satellites into space; an armada of computers to set them on the right orbit to emit on the correct frequencies and encrypt communications using sophisticated digital tools; legions of super calculators to analyse the deluge of data; and, to direct this data in real time, a planetary mesh of underwater cables, a maze of overhead and underground electricity networks, millions of computer terminals, countless data-storage centres, and billions of tablets, smartphones, and other connected devices with batteries that need to be recharged. Thus, the supposedly virtuous shift towards the age of dematerialisation is nothing more than an outright ruse, for there is no end to its physical impact.[29] Feeding this digital leviathan will require coal-fired, oil-fired, and nuclear power plants, windfarms, solar farms, and smart grids — all infrastructures that rely on rare metals.

Yet not a word about this is uttered by Jeremy Rifkin. So I reached out to the illustrious thinker to discuss the material reality of this supposed invisibility and the paradox of green energy. Repeatedly, I contacted the Foundation of Economic Trends, through which he offers himself as a speaker and consultant. I sent letters — via email — requesting clarification on this contradiction. I also suggested a brief meeting with Mr Rifkin during one of his trips to France, and while we were in the Washington suburb where his offices are located.

My questions were left unanswered. Perhaps on account of the fatal error underlying the energy and digital transition: they were not ground-truthed. It's all very well that green tech is thought up in the mind of a scientific researcher, finds real-world applications through the efforts of a persevering entrepreneur, benefits from attractive tax regimes and lenient regulations, and has the backing of plucky investors and benevolent business angels. But we forget that all green technology begins prosaically as a gash in the Earth's crust. This new demand on the planet replaces our dependence on oil with an addiction to rare metals. We are offsetting deprivation with excess — a bit like drug addicts weaning themselves off cocaine by sinking into heroin. Instead of addressing the challenge of humanity's impact on ecosystems, we are displacing it, and the rate at which we are subjugating today's environmental perils could very well have dire ecological consequences.

The broken promises of recycling

Can we moderate our consumption by recycling rare metals on a large scale to mitigate their impact and that of their extraction?

The idea is so appealing that the Japanese have already started to put it to work. It's an autumn afternoon in 2011, and I am in the suburb of Adachi, in the north of Tokyo. The tranquil atmosphere is suddenly disrupted by a commotion of blue-bin lorries. One of their drivers, Masaki Nakamura, is collecting electronic waste — e-waste — and piles up all manner of old video game consoles, mobile phones, and television screens in the back of his lorry. At the end of his round, Mr Nakamura takes his bounty to the dump of the Kaname Kogyo recycling company nearby. We find its CEO, Matsuura Yoshitaka, complete in a dark suit and tie, clambering over the mounds of home appliances that are being meticulously sorted by his teams. 'Nowadays people throw out appliances without much afterthought,' Yoshitaka tells me amid the crash of metal being moved from tip to tip. 'Yet they're full of rare metals!'

Clearly, globalisation has hurtled us into an unprecedented era that has made Western countries so prosperous that even our waste makes us rich — be it food, household, industrial, nuclear, or electronic. We have gone from a world in which our grandparents battled with daily deprivations to a civilisation that is at a loss about what to do with its staggering surpluses. After having to decide what to consume, we are faced with the conundrum of what to do with what we've already consumed.[30] Take scrap metal: every year, in France, every person produces an average of up to 28 kilograms of electronic waste.[31] Worldwide, the volume of this waste in 2021 was estimated at 52 million tonnes — an annual increase of 3 to 4 per cent.[32]

Until now, industry has contented itself with recycling the big metals, and has done so successfully: around 85 per cent of gold, 50 per cent of silver, and 45 per cent of copper and aluminum is reprocessed.[33] But until recently no one has taken a real interest in the smaller, hidden metals. This is where Japan has broken new ground by recognising the sheer quantity of rare-earth metals contained in the thousands of 'urban mines' (e-waste dumps) littering its territory. Around 5 million mobile phones are thrown away every year in Japan, each containing a few tenths of a gram of rare metals. The potential annual output of this urban mine would be more than a tonne of neodymium and several hundred kilograms of dysprosium.[34]

This realisation is the driving force behind one of the most innovative circular economies for e-waste. (See Appendix 9 on the lifecycle of metals.) Every year, collection drives are held across Japan to return 500,000 tonnes of digital waste to the consumption circuit.[35] The campaign is so popular that it has earned the endorsement of Japan's favourite virtual celebrities, including Hatsune Miku. Against a background of manga *Keitaï* (mobile telephones), the scantily clad anime popstar squeaks in electronic tones about the goldmine that is the country's waste.

But the campaign is not enough, and Tokyo has also invested hundreds of millions of dollars in scientific research to find substitute metals[36] and to reduce the rare-earth–metal content of its magnets.[37]

Other Western countries have begun to follow suit. The US army, for instance, a formidable consumer of rare metals, has out-of-service aircraft on the outskirts of Tucson, Arizona, as far as the eye can see, each one concealing tonnes of rare-earth metals that US army generals are unable to extract or reuse.[38]

Worse still is the fact that as the world's most powerful army withdrew from Afghanistan, it is believed to have left behind $6 billion worth of military equipment crammed with magnets that its enemies could dispose of as they pleased.[39] Many in the United States have recognised just how big a challenge it is, and have suggested giving soldiers instructions on how to retrieve components containing rare-earth metals before demobilising.

For industry, it's another matter entirely, for the circular economy has completely upended traditional supply chains. In addition to having to know where their raw materials are sourced, manufacturers must now also be able to trace the users of their product. Companies such as Apple and H&M, which of course know where their rare-earth minerals and cotton bales are coming from, now have to trace the billions of iPhones and old jeans scattered across the globe. In other words, the roles of sender and recipient have been switched.[40]

Doing things in reverse for the same outcome is, for many, a Copernican revolution. But it is also a way to increase the portion of recycled metals in the supply chain. It could also give us a glimpse into the future of rare metals: a world in which the mining powerhouses are not the countries with the richest mineral deposits, but those with the most bountiful waste dumps; where treasure maps of the world's biggest scrap mountains will be drawn, and some debris will be ranked 'world class' in the same way as some of today's mineral deposits. Our waste dumps will be coveted goldmines.

Accordingly, Japan hardly mines any rare metals from its soils. Its pre-eminence in this circular economy could also turn it into a powerful exporter of those retrieved metals on which other countries depend. Thus, the geopolitics of recycling would be born — at least, this is what Japan believes. It's also not hard to imagine the technological advances that would make production more ecological, reducing the need to mine and to export old television sets to e-waste dumps in Ghana and Nigeria.

While this ambition looks good on paper, it is incredibly complex to implement. Unlike traditional metals such as iron, silver, and aluminium, rare metals are not used in their pure state in green technologies. Rather, the manufacturers in the energy and digital transition are increasingly partial to alloys, for the properties of several metals combined into composites are far more powerful than those of one metal on its own. For example, the combination of iron and carbon gives us steel, without which most skyscrapers would not be standing. The fuselage of the Airbus A380 is in part composed of GLARE (Glass Laminate Aluminium Reinforced Epoxy), a robust fibre–metal laminate with an aluminium alloy that lightens the aircraft. And the magnets contained in certain wind turbine and electric vehicle motors are a medley of iron, boron, and rare-earth metals that enhance performance.

Today, there is a profusion of new materials — translucent concrete, paper bricks, insulating gels, and reinforced wood — that transform the properties of the original material. These alloys are so promising that green technologies will increasingly become dependent on them. But, as the name suggests, alloys need to be 'dealloyed' to be recycled.

Numerous technologies for dealloying rare metals already exist. I learnt about one of them at Toru Okabe's lab at the University of Tokyo in 2011. Amid a tangle of cables, pipes, and thermometers, the researcher demonstrated his latest invention: a high-temperature autoclave that uses salt from the high-altitude salt flats of Bolivia. 'We use the salt to selectively leach the rare-earth metals from other metals,' he told me.

Clearly, recycling an alloy is anything but straightforward. Understanding it takes us back to our bread metaphor. To save the loaf of bread left over at the end of the day, our baker will try to separate its ingredients — an insanely complex, time-consuming, and energy-intensive endeavour. The process is no less difficult for the rare metal magnets found in wind turbines, electric vehicles, and smartphones: manufacturers have to use time-consuming and costly techniques involving chemicals and electricity to separate rare-earth metals from other metals.

If alloying is like marriage, recycling is like divorce: it comes at a price. 'The technology I'm showing you has a lot of potential, but it's far from financially viable,' Toru Okabe admitted. Thus, the rare metals in Japan's waste dumps are hidden treasures that no economic model today can retrieve. It is the prohibitive cost of recovering rare metals — a cost that currently exceeds their value — that is holding industry back.

There is a bright spot: if the price of extracted materials is high, the price of recycled metals is competitive. However, this puts recyclers at the mercy of the market and its inherent unpredictability. It also hinders long-term investments because of the constant rise and fall in the price of mined metals, making recycled material competitive one day, but too expensive the next.[41]

It is difficult, therefore, for manufacturers to recycle large quantities of rare metals. Why rummage through e-waste dumps when, often, it is infinitely cheaper to go straight to the source? This could account for the sluggish investment in recycling processes and facilities. More and more of some of the base metals are being recycled, such as copper (55 per cent), aluminium (32 per cent), tantalum (40 per cent), and vanadium (27 per cent); however, recycling rates are still very low for the vast majority of metals, such as lithium and rare earths (between 0 and 10 per cent), which are essential for electric transport. (See Appendix 10 for a summary table of rare metals recycling rates.) There are major efforts underway to develop new recycling processes and industries, but we shouldn't expect to see significant volumes before the 2035–2050 period.[42]

The conclusion? The volumes of recycled metals will still fall structurally short of demand. Even recycling nearly 100 per cent of lead has not been enough to stop its mining and extraction, because of perpetually growing demand.[43] It would seem that hell is well and truly paved with good intentions.

Return to sender

Despite this, manufacturers agree that rare metal recycling can be profitable by accumulating enough volume to create economies of scale. The problem is that in addition to being difficult to recycle, these metals don't stay in one place.

In the state of New Jersey, Newark Bay offers more than just views of the handsome towers of Manhattan. It hosts myriad US companies specialising in recycling e-waste — or, rather, exporting it. Their proximity to the nearby coastal ports is not lost on Lauren Roman, an activist from the US nonprofit organisation Basel Action Network. For years, she has been scouring New Jersey, photographing the tracking numbers on containers loaded with IT waste. With these numbers, she then tracks their journey around the world.

The overwhelming majority of recycling companies are obliged to process e-waste in the countries of its collection. This is the very thrust of the United Nations Basel Convention.[44] Adopted in 1989, the convention prohibits the movement of waste that is considered hazardous, on account of its potentially toxic heavy metal content, to countries with lower environmental standards.[45] To date, 186 countries and the European Union are parties to the convention, but a handful of countries — including the United States — have refused to ratify it. This emboldens US recycling companies to export unusable e-waste. After years of investigation, Lauren Roman confirms without a shadow of a doubt that 80 per cent of e-waste produced in the United States is sent to Asia.

This situation is not unique to the US. Japanese recycling companies also export their e-waste to China, despite being a party to the Basel Convention. One murky yet well-established line of business is *Kaitori* (vendors selling second-hand goods): waste buyback services that flout international regulations by loading used equipment into containers dubiously labelled 'humanitarian aid'.

Europe isn't doing any better. Countless 'second-hand' vehicles full of rare-earth metals leave the docks of Amsterdam.[46] Adding to this loss is 50 per cent of used catalytic converters, a monumental stock of wind turbine batteries, over 50 per cent of used electronic cards, and a million tonnes of copper per year. Despite Europol classifying, from 2013, the illicit trade of waste as one of the biggest threats to the environment, progress is slow.[47] In fact, only in early 2023 did the European Parliament and EU governments begin negotiating new procedures for and restrictions on waste exports.[48]

China is a major destination for e-waste, as the country's labour is so much less expensive than in Europe.[49] As for magnets that cannot be recycled at an acceptable price, my Chinese contacts tell me that they are stored pending an economically viable solution — a claim I was unable to verify.

What can we conclude?

- 'Green' technologies require the use of rare minerals whose mining is anything but clean. Heavy metal discharges, acid rain, and contaminated water sources — it borders on being an environmental disaster. Put simply, clean energy is a dirty affair. Yet we feign ignorance because we refuse to take stock of the end-to-end production cycle of wind turbines and solar panels. 'It's no longer enough to look at the green, sound, and non-polluting end product,' stresses Chinese environmental activist Ma Jun. 'We need to take a good look at whether the sourcing and industrial manufacture of the components making up the technology are environmentally friendly.'

- These very energies — deemed 'renewable' because they use energy sources (such as sunshine, wind, and tides) to which we have unlimited access — rely on mining resources that are not renewable. The wealth of minerals in the Earth's crust is finite, and the billions of years needed for their formation is no match for the exponential growth of our needs. I will revisit this point.

- In reality, these energies — still qualified as 'green' or 'carbon-free' because they help us wean ourselves off fossil fuels — depend on activities that produce greenhouse gases. It takes prodigious quantities of electricity from power plants to operate a mine, refine minerals, and transfer them to a manufacturer to be used in a wind turbine or solar panel. The result is tragically ironic: the pollution no longer produced in urban areas thanks to electric cars has merely been displaced to the areas where the resources needed for this very technology are mined. The energy and digital transition therefore belong to the wealthy. The most affluent areas are relieved of pollution by dumping its impacts on the most destitute and remote areas. But what can we do if we're not aware there is a problem in the first place? This is where our energy model is particularly pernicious: in contrast to the carbon economy, whose pollution is undeniable, the new green economy hides behind virtuous claims of responsibility for the sake of future generations.

- The technologies lauded by many in ecological circles as the ticket out of nuclear depend on resources (rare-earth metals and tantalum) that release radiation when mined. While rare metals are not radioactive in themselves, the process of separating them from other radioactive minerals — such as thorium and uranium — to which they are naturally associated produces radiation in amounts that cannot be overlooked. Expert accounts put

the ambient radioactivity of the toxic reservoir in
Baotou and at the bottom of the Bayan Obo mining
district at twice that recorded in Chernobyl today.[50]
And despite the low level of radioactivity of normal
mining conditions, International Atomic Energy
Agency standards still require the generated waste to
be isolated for several hundred years.[51]

- To accelerate the energy and digital transition, a few
  well-meaning souls would like to apply the concept
  of 'local loops' — used for the management of food
  products — to energy distribution. Eco-suburbs,
  such as Vauban in the German city of Freiburg im
  Breisgau, pride themselves on using only clean energy
  that is increasingly local. But has anyone thought to
  add the millions of kilometres travelled by the rare
  metals, without which these areas would not exist?
  'Eco ideas like these work in theory. But in practice
  they are completely misguided,' a concerned expert
  tells me.[52]

- Some of the green technologies underpinning our
  modest consumption ideals ultimately require more
  raw materials to be mined than do older technologies.
  A report by the World Bank states that 'a green
  technology future is materially intensive and, if not
  properly managed, could bely the efforts ... of meeting
  climate and related Sustainable Development
  Goals'.[53] Denying this reality could very well lead us
  to go against the goals of the Paris Accords. It could

also result in a shortage of exploitable resources: the global population of 7.5 billion individuals over the next three decades will consume more metals than the 2,500 generations before us.

- Lastly, recycling the rare metals at the heart of our greener world is not as eco-friendly as we are led to believe. Its environmental footprint could even increase as alloys become more complex and involve more materials in greater proportions. Industrial companies in the energy and digital transition will have to contend with an immovable contradiction: their quest for a more sustainable world could effectively stifle the emergence of new, modest consumption models hinged on the principles of the circular economy. Perhaps future generations will look back and say: 'Our twenty-first century ancestors? They're the ones who took metals out of one hole and put them in another!'

These observations may seem obvious to those in the business of commodities. But to the vast majority of us, they are so counterintuitive that it may be many years before we fully apprehend and admit to them. Come that day, we will pick apart decades of myth and illusion. We will listen more intently to the warnings of the Cassandras. Like Carlos Tavares, the head of France's biggest car manufacturer, PSA, who, at the Frankfurt International Motor Show in September 2017, warned of the harmful effects of electromobility on the environment: 'If we are instructed to make electric vehicles, authorities and administrations must assume the scientific responsibility of this choice. Because I don't want to find that in 20 or 30 years, we've fallen short in aspects such as battery recycling, how we make use of the planet's scarce resources, or the electromagnetic emissions we generate when recharging batteries.[54]

Perhaps we can expect an 'electricgate' that, like the 'dieselgate' scandal, will result in legal proceedings on a global scale. We will wonder how we could have turned a blind eye to the mounting evidence. We will admit that the consensus between economic and political circles, and backed by numerous environmental groups, drowned out any arguments to the contrary. We may even reach the conclusion that nuclear energy is in fact less harmful than the technologies we wanted to replace it with, and that we'd be hard pressed to do without it in our energy mix.

We will need to design new technologies, which some will undoubtedly qualify as 'miracle technologies', to rectify the immense problems this heedless march towards a greener world will have caused.

And we will not be able to lay the blame on the Chinese, Congolese, or Kazakhs alone. Westerners have brought this situation on themselves by knowingly allowing the most irresponsible countries to flood the rest of the world with dirty metals.

CHAPTER THREE
Delocalised pollution

RATHER THAN TAKE THE LEAD IN RARE METALS, THE WEST chose to shift its production and accompanying pollution to poor countries willing to sacrifice their environment for financial gain. I decided to see this for myself. One morning in 2011, leaving the buzz of Las Vegas, I headed south-west along Interstate 15 — that straight asphalt strip cutting across the vast expanses of Nevada and California. Two hours later, I found myself looking down a quarry in the midst of a corroded industrial wasteland, a tired star-spangled banner ruffling overhead. Until 1990, US mining company Molycorp operated in this depression of rocks and shrubs that is the Mountain Pass mine. It is also the biggest rare-earth mine on the planet.

When the US dominated the rare-earth metals market
Contrary to popular belief, rare metals reserves are not concentrated in the world's most active mining countries, such

as China, Kazakhstan, Indonesia, or South Africa.[1] They are
spread across the planet in varying degrees of concentration,
making them both rare and not rare. The most strategic of
these mineral reserves — rare earths — are found in a dozen or
so countries.[2] A report by the French parliament states: '[B]efore
1965, extraction took place in South Africa, Brazil, and India;
but total production was marginal: fewer than 10,000 tonnes
per year.'[3] Between 1965 and 1985 there was a second phase, during
which the United States was the global leader in rare-earth
metal mining. The report continues: 'Although they did not
have the monopoly, they were clearly dominant, with quantities
as high as 50,000 tonnes per year.'[4] Mountain Pass was the mine
that supplied these resources.

But the environmental damage caused by Molycorp's
operations began to have a serious impact on the surrounding
ecosystems. Clearly, the situation was an embarrassment for group
management, for our request to visit the quarry was flatly refused.

So, in an act of desperation, I contacted a light aircraft
rental company and informed Molycorp that since I wasn't
allowed to walk through the facility, I would enter their airspace
the following day at noon. Accompanying me that day was
John Hadder, executive director of a very vocal environmental
nonprofit organisation in the region, Great Basin Resource
Watch (GBRW).

The next day, from the tarmac of a Las Vegas aerodrome
surrounded by mauve-tinted mountains, our rickety plane was
catapulted to altitude and, within minutes, we could make out
the Molycorp mine in the middle of the mineral landscape.
Soon afterwards we were flying over the rocks of the quarry and
spiralling down to a body of water.

The most instructive part of the flight extended some 20 kilometres away from the excavation itself: a circular decant pond spread over several hundred metres in the heart of the desert. 'When the mine was in operation, all the discharged water was channelled to the pond,' John tells me. 'This contaminated water is still seeping into the groundwater.'

Back on terra firma, in the shadow of a hangar next to the runway, the document John Haddar showed me was edifying: a map of Mountain Pass based on intelligence gathered between 1984 and 1998 by the United States Environmental Protection Agency. GBRW must have requested the declassification of this information to allow them to map the ecological damage caused by Mountain Pass's operations in the Mojave Desert.

What struck me was the succession of numbers reproduced on the map between the excavations and the decant pond. 'The water polluted by the ore processing was pumped out of the quarry and evacuated into this big pond,' Hadder explained. Directing the billions of litres of wastewater was a pipeline. 'The numbers shown here along the pipeline show where there were ruptures or leaks.' Over fifteen years, some sixty spills occurred. 'The worst was in 1992: a leak of one and a half million litres! In total, nearly four million litres of wastewater have spilled into the desert.'

This environmental damage hit local communities hard. The soil was contaminated with a noxious blend of uranium, manganese, strontium, cerium, barium, thallium, arsenic, and lead.[5] Polluted sand and contaminated groundwater fouled the Mojave Desert for miles around. 'After a series of lawsuits, the company was forced to tackle the environmental issues head-on,' John told me. Molycorp was fined heftily.

Once, Molycorp was even paid a visit by armed federal agents, complete with shields and bulletproof vests, to notify the company of yet another violation of California's environmental protection regulations. And to protect the nearby habitats of the desert tortoise, the mine's 300 employees were ordered to attend training to learn about the reptile, and were prohibited from coming within 30 metres of them. At the end of the 1990s, fearing new spills and faced with the astronomical cost of modernising their facilities, Molycorp began to reassess the continuation of its operations at Mountain Pass.

Meanwhile, a China that was thirsty for economic growth saw in this challenging period for Western mining companies an opportunity to become a dominant player in the rare metals market. This was an ambition not without substance, for its mines in Baotou (see Chapter One) represented nearly 40 per cent of the world's rare metal reserves.[6] To accelerate the shift of mining production from West to East, China used — and continues to use — formidable cunning that is captured in just one word: dumping. It engaged in trade dumping by slashing production costs; and environmental dumping because, as environmental activist Ma Jun explains, 'production costs do not factor in the cost of repairing the environmental damage'. And has China done anything at all to even paper over the damage?

Naturally, this dual-dumping strategy dragged down Beijing's prices, and by 2002 a kilogram of rare-earth metals produced in China had an average cost price of US$2.80 — half that in the US.[7] Molycorp did not stand a chance against this ruthless competition, and shut down its operations at Mountain Pass, running down its stocks until finally closing the mine in 2002.[8]

An environmentally ethical approach would have been to set aside the quest for financial gain by subsidising, albeit at a loss, rare metals mining in Western countries where ecological responsibility held currency. This was a responsibility that could have fallen to Australia. In 2001, a wealthy Australian businessman, Nicholas Curtis (who has since become chairman of the board of directors of the mining company Northern Minerals) acquired Mount Weld — one of the biggest rare-earth deposits on the planet, located in the state of Western Australia. 'Curtis believed these materials would become highly strategic given the diversity of their applications, be it for cars, phosphors, or televisions. He founded the junior mining house Lynas to extract cerium, lanthanum, and neodymium from the Mount Weld mine,' said an expert on the matter.[9]

But the businessman had to face the fact that Lynas stood no chance of competing against the rock-bottom prices offered by the Chinese. The financial crisis in 2008 further jeopardised the opening of the mine.[10] And all the while, countries in the West left the 'sorcerer's apprentice' of rare metals to his own devices. 'We in the West knew full well at what cost we were accessing rare-earth metals that were more or less clean, and which posed no risk to their future generations,' said another French expert. 'But we preferred to turn a blind eye to what was happening in China.'[11]

The Americans are not the only ones to have washed their hands of the affair. They share this responsibility with the French.

Hot-air balloons, adventure, and rare earths: the Rhône-Poulenc saga

France, a leading supplier of resources for the third industrial revolution? To jog our memory, let's take a look back at one of the shining periods of the humble television. In France, many will recall those Saturday nights in the 1980s at around 10.00 pm when everyone gathered around the box to watch *Ushuaïha*. The show's presenter was environmentalist Nicolas Hulot, France's answer to David Attenborough.[12] Viewers were transported to far-flung places to encounter little-known peoples, discover exotic animals that would have had Rudyard Kipling green with envy, and drift silently over epic landscapes in the helium-inflated envelope of a hot-air balloon. Broadcaster TF1 sold viewers the dream — while also selling off the little attention they had left. At the very bottom of the screen could be seen the rectangular logo of the program's official sponsor, French chemical company Rhône-Poulenc, together with the tagline: 'Welcome to the world of adventure, human feats, and exploits!'

Before its chemical arm became Rhodia in 1998 and it merged with Belgian group Solvay in 2011, Rhône-Poulenc was one of the two biggest rare metals chemical companies in the world. In the 1980s, its factory in La Rochelle, in the west of France, purified between 8,000 and 10,000 tonnes of rare earths every year — that is, 50 per cent of their global supply of 'separated' rare earths.[13]

This merits repeating: half of the most strategic rare metals — the resources of the future that would shape the energy and digital transition — were beneficiated in France. The country possessed unparalleled chemical knowledge coupled with superior commercial acumen. So much so, in fact, that French intelligence agencies monitored access at the plant just in case a Russian or Chinese guest on a working visit decided to indulge in a bit of industrial espionage.

The La Rochelle plant sits on a 40-hectare site along the coast west of the city. It is still in operation, and I was able to visit it in 2011. In the shadow of colossal warehouses, rare earths are separated and calcined in high-temperature furnaces to produce oxides (powders) that are packaged for the market. In the adjoining warehouses are row upon row of tightly packed bundles stamped with names we are now familiar with: cerium, dysprosium, neodymium, terbium, and so on. There is also a research and development (R&D) laboratory, but the activity of the refinery itself is a shadow of its heyday thirty years ago.[14]

It's no secret that extracting and refining rare earths produces high amounts of pollution. This is due to their natural association with radioactive elements, such as thorium and uranium. During separation at the La Rochelle plant, radon was released, albeit in minute quantities. Former employees of the French industrial company insist that inhaling the weakly radioactive gas has never had any impact on worker health. Yet '[t]here is no such thing as rare-earth minerals that are not radioactive'. And that's coming from a former director at Rhône-Poulenc.[15]

The few dozen tonnes of uranium that Rhône-Poulenc separated every year were sold to the national electricity utility EDF for its nuclear power stations. The thousands of tonnes of thorium were, and still are, stored at the plant in the hope that they could one day be used as fuel for a new generation of cleaner nuclear power stations.[16] The effluent by-product of separating the minerals went to a wastewater treatment plant before being discharged into the sea via an outfall pipe located on the shore of Port Neuf, at the end of the bay of La Rochelle.[17] The liquid waste had a high concentration of iron, zirconium, aluminium, silicon, magnesium residues, and other impurities. In the 1980s, there were numerous incidents of untreated sludge 'escaping' from the wastewater treatment plant to be decanted directly into the sea.

Did these discharges contain radioactive thorium? One former Rhône-Poulenc employee says no: all thorium was stored upstream of the treatment plant — that is, before the evacuation of effluents. He added that any radioactivity would only have emanated from the radium found in thorium and uranium. We should also point out that Rhône-Poulenc went to great lengths to mitigate the inevitable pollution of its operations. But non-governmental organisations (NGOs) believe otherwise, and estimate that since 1947 the plant has evacuated some 10,000 tonnes of radioactive waste into the ocean.

In 1985, a prefectural order toughened environmental regulations and banned Rhône-Poulenc from discharging any effluents through pipelines not submerged at high tide, when waste could be swept away by the sea currents. The order also capped effluent discharges. Nevertheless, at the request of a local association, Les Verts Poitou-Charentes, the Commission for Independent Research and Information on Radioactivity (CRIIRAD) conducted several inspections of the facility from 1987. Contrary to official reports, they found that the outflow pipe was in service at both high and low tide. Samples taken in the immediate surroundings of the evacuation channel confirmed 'extensive discharges and an accumulation of thorium and its progeny in the marine environment', as stated in a letter from the association to La Rochelle's MP and mayor, Michel Crépeau (1930–1999). The radioactivity near the outfall pipe was at 1,000 counts per second, or 100 times the local average. 'It was a very serious affair,' commented an engineer at CRIIRAD thirty years later.[18]

Not that this ruffled any feathers with the local authorities. 'Everyone knew [about the radioactivity], starting with the mayor Michel Crépeau. It was regularly reported on by the local press, but nothing was done about it,' says Hélène Crié, a reporter covering the events at the time for French daily newspaper *Libération*.[19] This contrasts with the position of Professor Pierre Pellerin, director of the French Radiation Protection Agency, which reports to the Ministry of Health: 'Every so often this business about radioactivity at La Rochelle crops up. If we keep up this nonsense, we are going to send everyone into a panic.'[20] Coming from the scientist accused of downplaying the fallout of the radioactive cloud that drifted over France after the Chernobyl disaster, one can't help but read this with circumspection.[21]

The environmental conscience of the French in the 1980s was a shadow of what it is today. But locals began to get hot under the collar. *Libération* wrote: 'Instructors at the local windsailing club worry that children sailing near that shore could accidentally swallow water or fall and hurt themselves in the contaminated mud. As for the bay as a whole, "because of the currents, it can take up to three weeks for a drop of water that's entered the harbour to be carried out."'[22] Committees were formed, and at public meetings community members could be heard chanting 'Rhône-Poulenc is an atomic timebomb!' and 'It's going to blow!'[23]

Member of Parliament Jean-Yves Le Déaut describes how he went to La Rochelle on two occasions to assess the situation — and was met by nearly 300 protesters with loudhailers. 'One gentleman came up to me and said: "We've been living a peaceful existence here, Mr Le Déaut. And now we are being exposed to radioactivity ... People here are starting to get scared."'[24] Jean-Paul Tognet, a former industrial and raw materials director of Rhône-Poulenc and Rhodia Rare Earths, recalls the 'growing criticism of Rhône-Poulenc's reputation by the media. Management wanted to pull the plug — the controversy almost shut down the La Rochelle plant.'[25]

From 1986 to 1998, Rhône-Poulenc was under the management of Jean-René Fourtou. In 1994, he changed tack entirely: 'I don't want to hear anything more about radioactivity. Buy whatever you need, but I will not allow a single radioactive product.'[26]

And with that, Pandora's box was opened. Rhône-Poulenc (which became Rhodia) found itself looking for foreign partners to carry out first-stage refinement. This is how, one fine day, the group asked Norway, India, and China whether they could make La Rochelle's non-radioactive products for them.

Were there errors in Rhône-Poulenc's mineral processing, and were the subsequent plant tours organised for the public too little, too late to redeem their lack of transparency? Did the residents of La Rochelle, with their limited understanding of radioactivity, exaggerate the risks to which they were actually exposed? And did the authorities raise suspicions by attempting to cover up certain information? No doubt everyone has their share of the blame. Either way, during this time other countries got ready to fill the void. 'In the early 1990s, the Norwegians supplied us with raw materials at high prices,' says Jean-Paul

Tognet. 'We should have kept the diversity of our supplies, but instead we stopped working with the Norwegians and entered into a long-term partnership with the Chinese, who were more competitive.'

Naturally, buying cheaper improved the French chemical company's bottom line. It was also profitable for its customers, eager to procure transformed rare earths at rock-bottom prices. It made perfect sense. Jean-Yves Dumousseau, who was sales director at the US chemical company Cytec at the time, explains: 'Obtaining rare earths from anywhere else would have been far more expensive than continuing to procure supplies from China at a quarter of the price! It's the same argument for electronics, jeans ... everything! I'm sorry to say it, but it was that simple.'[27]

Meanwhile, thousands of kilometres away from La Rochelle, China claimed the monopoly on the production — from extraction to refining — of rare-earth oxides, of which it produces over 100,000 tonnes annually. As for working conditions, '[t]here were no checks on the separation units or even safety procedures; you'd find the guys in the refineries doing electrolysis at 700 degrees without hardhats, wearing flip-flops and shorts! It was outrageous!' says Dumousseau.[28] Another source put it more bluntly: 'It was a mess, and no one wanted to deal with it.'

A new world order

The Rhône-Poulenc saga brings to mind those who despair of where the world is headed. To find a sense of order in this 'chaos', they latch on to clean divides: countries of the north and countries of the south; countries of the east and countries of the west (once separated by the Berlin Wall); emerging countries and developed countries; the Orient and the Occident; the free world versus the axis of evil; 'Old Europe' against the vibrant 'New World'; and so on. But there may be an even deeper divide that began thirty years ago — one that tells the story of the world of our making. It did not follow a treaty in Versailles, a congress in Vienna, or a conference in Yalta. Rather, it is an industrial order that formed organically between China and the West (to polarise once more).

It was first formally acknowledged in the infamous 'Summers Memo', an internal document at the World Bank signed in 1991 by its chief economist, Lawrence Summers. He purportedly recommended that developed economies export their polluting industries to poor countries, and especially to 'underpopulated countries in Africa [that are] vastly under-polluted', as 'the economic logic ... is impeccable'.[29] Anxious to explain himself after the memo was leaked, Summers naturally pleaded intentional sarcasm. Yet his comments are perfectly aligned with the reality: as our societies strive towards 'zero risk', all kinds of industrial activities have steadily been pushed out of Europe and the rest of the Western world.[30]

Consider the example of REACH (registration, evaluation, authorisation, and restriction of chemicals) — the European regulation aimed at minimising the sanitary risks associated with over 30,000 chemicals found in consumer goods.[31] It has vastly improved the quality of life of the European Union's 450 million citizens, and especially that of industrial workers.[32] But these environmental safety requirements have also taken the wind out of the sails of Western industry, which has had to suspend the production of myriad chemical substances that are now banned.[33] Now, others are free to take over production. And become Europe's suppliers.

The same logic applies to green technologies. In the last two decades of the twentieth century, the workload of the future energy and digital transition fell naturally between China, which did the dirty work of manufacturing green-tech components, and the West, which could then buy the pristine product while flaunting its sound ecological practices. Thus, the world was ordered as Larry Summers intended: between the dirty and those who pretend to be clean.

This is precisely the takeaway of the Mountain Pass and La Rochelle sagas: by moving the sourcing of its rare metals to China, the West chose to relocate its pollution.[34] We have knowingly and patiently created a system that allows us to move our 'filth' as far away as possible, and the Chinese — far from pinching their noses — have welcomed the initiative with open arms. As magnanimously put by a Canadian rare metals industrialist: 'We can thank them for the environmental damage they have endured to produce these metals in our place.'[35]

Rare earths are only the most obvious example of this systemic offshoring. In fact, almost the entire extractive industry has gradually moved from the West to countries of the South over the last few decades. (See Appendix II for the distribution of world mining from 1850 to 2030.) Europe, which accounted for more than 60 per cent of global mining production in the mid-nineteenth century, now accounts for no more than 3 per cent. The same is true of the United States, whose share has fallen from 40 per cent post–Second World War to less than 5 per cent today. While Canada and Australia have resisted this shift in production, it is primarily the developing countries of

Africa, South America, and central Asia that have steadily taken the lead.[36]

At this point, we should steer clear of any anti-capitalistic arguments: the countries on our radar have adopted this system of their own volition by deliberately shaping their economies to generate massively inflated profits. Many have done well on the progress afforded by the globalisation of the markets. Except now the Chinese have realised something else. A Chinese academic explains: 'We are praised for moulding to the Western-ruled world order of the time. But China also suffered. And in terms of cost-effectiveness, I am not wholly convinced of the advantages in our favour.'[37] Under this arrangement, Beijing has effectively laundered dirty minerals. Concealing the dubious origins of metals in China has given green and digital technologies the shining reputation they enjoy. This could very well be the most stunning greenwashing operation in history.

Naturally, businesses in the West are complicit. 'They couldn't care less about the conditions under which the minerals are extracted and refined,' a European industrialist told me. 'All that matters is having the lowest price possible.' Too often, they are also complicit in our ignorance of the human dramas behind the scenes of the energy and digital transition. An impact report published by Tesla in 2022 addressed the critical metals used for batteries, but made no mention of the rare earths needed for the on-board electronics and electric motor in some of its models.[38]

Consumers could have led the resistance: their choice to buy or boycott has the power to redirect the market and change its practices. They had the information: numerous documentaries have exposed the distressing environmental and social impact of electronic goods, as have reports by NGOs and supranational organisations.[39] Consumers would have needed to put pressure on manufacturers to design more ecological products, such as the repairable Fairphone,[40] and to use their votes to pressure their governments to beef up feeble anti-obsolescence regulations. But consumers would hear nothing of it: a connected planet was better than a clean one.

But there is hope. In France, the anti-waste law for a circular economy was adopted in 2021 to eradicate disposable consumer products and the programmed obsolescence of equipment, and to extend the lifespan of such equipment.[41] As early as 2015, the European Commission introduced measures to accelerate the circular economy, and is regularly tightening up anti-waste regulations.[42] Such citizen-led initiatives to repair domestic appliances have mushroomed.

Today, Europe and the US want their whites whiter than white. In 2019, the former adopted the European Green Deal, encompassing numerous climate, energy, and environmental objectives, with a view to achieving carbon neutrality by 2050.[43] In 2022, the US adopted the Inflation Reduction Act to promote green energy, industrial reshoring, and e-mobility, and has set extremely ambitious targets for reducing its greenhouse gas emissions.[44] Yet how could we possibly hope to stay on course if its polluting industries came home? Environmental responsibility has already been transferred, and reversing this transaction will be a long and difficult task.[45] Therefore, is the West — brandishing its self-declared legitimacy — truly in a position to lead talks on the fight against global warming? Shouldn't COP 21 have been held in Beijing, Kinshasa, or Astana rather than in Paris? And is France really in a position to urge, as President Macron did in June 2017, that we 'Make Our Planet Great Again'?

The truth is that we are no better than companies that boast colossal earnings to their shareholders — while hiding a mountain of debt in an obscure subsidiary in the Caribbean. These 'off-balance-sheet' transactions that use fraudulent accounting practices have led to the conviction of countless company directors. We glorify our modern ecological legislation — while shipping out our electronic scrap to toxic waste dumps in Ghana, exporting our radioactive waste to the ends of Siberia, and outsourcing rare metals mining around the world, to make a deadweight loss look like a net profit.

The illusion of a new era of opulence

The zeitgeist of the 1990s played an important role in this reshuffling of roles. At the start of the decade, George Bush and Margaret Thatcher gave wing to the expression 'peace dividend'. The fall in military spending at the end of the Cold War gave rise to a new era of peace and economic prosperity. One can recall the feeling of optimism — euphoria even — that reigned at the time. With the end of the nuclear arms race came demilitarisation, and the states that had built up stockpiles of rare metals in preparation for an armed conflict were left wondering whether they were worth holding on to.[46]

A strategic stockpile is like a savings account: when the forecast looks gloomy, we tuck some money away for that rainy day. Naturally, in the bright days of optimism, we dip into our savings and enjoy the sunshine. So, in the 1990s, we witnessed a global sell-off of strategic stockpiles.[47] In France, the platinum and palladium stockpiles deposited in the safes of Banque de France were quickly sold off by both left- and right-wing governments. The dozens of billions of dollars' worth of lithium and beryllium rare earths in the US went the same way.[48]

This cornucopia largely originated in the former USSR and its satellite states. Palladium stockpiles were sold discreetly and in record numbers via Zurich. As a former asset manager relates: 'It started with the banks UBS and Credit Suisse, working for end-consumers in the jewellery sector, for instance. Palladium was also bought by trade majors like Glencore and Trafigura.'[49] Meanwhile, Beijing pursued a quite different strategy of building its reserves while buying up a handsome portion of the market inventory.

The sudden superabundance of raw materials on offer sent prices into a long-lasting decline that gave the impression of limitless availability — an illusion of an era of abundance given substance by the breaking down of international commodity trade barriers. In the mining sector, businesses cared only about buying metals as cheaply as possible.

It was only a matter of time before this elusive opulence diverted industrial companies from their duty of knowing the origin of the resources they relied on and managing their supply. This is precisely what happened in the lumber industry. In Europe, many producers and artisans no longer know the exact origins of their materials. As production lines have scattered and the distances travelled by timber have grown longer, parquet floor layers, carpenters, and the like have become disconnected from the substance of their trade. The problem is that the day China suddenly decides to snap up French oak, for example, no one will know where to find new suppliers.[50]

It's the same story for perfumes. From the 1950s, globalisation and the low cost of labour led to perfume-makers neglecting their flower farms in Grasse, in the south of France, in favour of less noble products from Egypt, India, or Bulgaria. 'Perfume was sold based on concepts rather than on the quality of ingredients,' a professional tells me.[31] It's a philosophy that the perfume sector is now reconsidering.

This is also more or less what happened in the industries that use rare metals. The assumption still exists that resources are available in any quantity and at any time. Supporting this utopia are the dogmas of 'just in time' and 'zero stock'. Taught in MBA programs the world over, and applied by big industrial groups, the two production management methods were established in 1962 by Taiichi Ohno, an engineer at the Japanese company Toyota (hence the term 'Toyotism'). 'Just in time' is about avoiding surplus stock by making the time between the manufacture of a product and its sale as short as possible, resulting in what is known as 'lean manufacturing'. Its corollary, 'zero stock', outsources the management of spare parts and components to a multitude of contractors — thereby transferring the risks associated with the delivery of raw materials to these third parties.

This complexification of supply chain logistics means that the exact origins of a raw material are not wholly transparent from one end of the supply chain to the other. Instead, participants can see no further than one level upstream and one level downstream. Surely this explains why, in a recent annual report, under the heading 'raw materials risk', Thales — a multinational whose rare metals–intensive electronics are used in aerospace, defence, security, and land transport — sanctimoniously states: 'Given the nature of its business, Thales uses few raw materials. The Group's exposure to raw materials risk is therefore negligible.'[32]

Hypocrisy? Ignorance? Either way, up until just recently, Toyotism helped relieve businesses of their responsibility with respect to 'metals risks'.[33]

An end to public policies on mineral sovereignty

We see this same insouciance and lack of foresight within governments, which have steadily put mining strategies on the backburner.[34] However, before the collapse of the communist bloc, this was not the case for France, as illustrated by the French Geological Survey (BRGM) — a public institution world-famous for its mining expertise. The BRGM even benefited from the oil crises of 1973 and 1979, which also served to familiarise French political leaders with the realities of resource scarcity. In 1978, as an initiative of the French minister of industry, André Giraud, the government launched the 'Metals Plan' — a vast mining stock-take aimed at boosting the BRGM's activities. Former employees nostalgically refer to this period as the 'golden age of BRGM', when it had a mining exploration program that enjoyed the active support of the government, especially in metropolitan France and French Guyana.[35] Its exploration department had 250 employees, and offered its services in francophone Africa, Portugal, and Quebec.[36]

In the 1990s, the Metals Plan came to an end, and the dynamic BRGM began to lose its lustre. By 2000, exploration activities were wrapped up, marking the start of what would be called 'the winter of French mining'.[37] 'To think we had gold, zinc, tungsten, antimony, and silver,' recalls a former employee, 'but investors were becoming increasingly rare.'[38] The mines that hadn't already been shut down were abandoned, creating their own lot of social dramas. 'The mining industry never made up a large part of French GDP,' says another former employee, 'but when you add up the few hundred employees per mine, you're left with a lot of people. At the end of the day, we lost an entire industry. Not to mention our mineral sovereignty, our capacity to supply our industries with our own minerals and metals.'

This underlying trend is common to all Western countries. One need only look at infographics on the history of global mining production. While Europe produced nearly 60 per cent of the world's heavy metals in 1850, its momentum steadily declined to produce no more than 3 per cent today. Mining production in the US hasn't fared any better: after peaking in the 1930s, accounting for close to 40 per cent of global production, it now represents around 5 per cent.[59]

Back in Europe, the three French White Papers on defence and national security produced in 1971, 1994, and 2008 made no reference to the supply of rare metals, despite how critical these materials are to military technology. Only in the 2013 publication did the term 'strategic materials' first appear.[60] It was of little interest to French intelligence agencies. 'The government never asked the agencies to take any action in this regard. I think the Directorate-General for External Security was light years away from such matters,' admits former intelligence director Alain Juillet. According to several corroborative sources, however, in the 1970s a former spymaster, Alain de Marolles, convinced the head of French intelligence, Alexandre de Marenches, to add eight strategic minerals to a metals supply risk list. De Marenches subsequently stepped down, de Marolles was 'honourably discharged', and their cause went no further.[61]

Three decades were enough to bring about an about-turn

in strategy. Up until then, a nation's power hinged largely on its ability to rely on its own vital resources — or, failing that, to do everything in its power to guarantee supply from outside its borders.

Consider the example of oil. In the UK at the start of the twentieth century, the First Lord of the Admiralty, Winston Churchill, made the decision to convert the Royal Navy from coal to oil. He also did this to secure oil supplies from the Middle East for his country. The British government acquired a controlling interest in the Anglo-Persian Oil Company, and crisscrossed Persia with gigantic marine pipelines.

As for the United States, when they realised after the Second World War that their own oil reserves would not be enough to meet their growing energy needs, they turned to the Kingdom of Saudi Arabia and its extraordinary crude oil reserves. The 'Quincy Pact', signed on 14 February 1945 between President Roosevelt and the Saudi king, Ibn Saud, gave Washington privileged access to Riyad's petroleum in exchange for military protection. France looked to Algeria and Gabon in the same way. When it came to food, however, Paris always managed to safeguard part of its sovereignty during World Trade Organization (WTO) talks by limiting the liberalisation of agricultural markets. Just as it safeguarded its civil nuclear program to maintain its energy sovereignty ...

For millennia, the fundamental rules of relying on one's own resources or securing a sustainable supply beyond one's borders have dictated every strategy for achieving energy autonomy. Yet to date, neither of these has been applied to rare metals. One could argue that the quantities at stake are inconsequential when compared to our gluttonous appetite for oil. But as we have by now discovered, these metals are as discreet as they are indispensable. Even though every person on the planet consumes as little as 35 grams of rare-earth metals a year, the world would run infinitely slower without them.[62] And yet few futurists have looked into how important these minute metals have become as a result of our technological choices since the 1970s. The policy of both demanding and claiming complete dependence on others is now widely accepted, whereas not that long ago it would have been considered utterly ruinous.

This was certainly before the market became so short-sighted, as this US expert says: 'Western countries no longer have long-term strategies, and rare metals are no exception.'[63] There is also the particular French context. France's mines and its abundant agricultural and fishing resources have made it less averse to dependence than other countries, like Japan, that have had no choice but to develop a trading culture and to find reliable supply routes to compensate for their lack of natural resources. And since France is not a country of traders either, it has not developed a culture of economic intelligence as far as the rare metals market is concerned. Says one specialist: 'The French DNA is not equipped for a situation of scarcity.'[64]

In short, the Western world honours the 'cargo cult' founded not too long ago in the Pacific Islands. Between the end of the nineteenth century and the 1940s, the mosaic of Melanesian peoples scattered predominantly across Papua New Guinea, the Fiji Islands, and New Caledonia came into sudden contact with Western societies — first with French and British colonisers seeking gain and conquest, and then with the US Army during the Pacific War. Both sets of fresh occupants of the region needed regular supplies of food and non-food items, so they built logistics networks that linked the spray of islands to the rest of the world.

Just imagine the stupefaction of those ancestral peoples upon seeing the arrival of boats and then planes, their holds loaded with treasures. And their amazement at how easily these goods appeared: all it took were radio operators to fire off their requests for medication, rations, and equipment to land on the fine-sand beaches or be parachuted down from the sky, as if by magic. Naturally, the Melanesians did not have the slightest notion of the industrial fabric woven behind these supplies. But because one apparently needed only to ask in order to receive, they began to imitate the Westerners, designing dummy radio devices, laying fake runways, watching, and waiting — for a very, very long time — for their needs to be met. The Westerners gave these rituals the name 'cargo cult'.[65]

In the twenty-first century, on the other side of the world, our societies — as informed and materialistic as they are — are giving in to a similar cult. The genius of logistics has stripped us of a fear that obsessed our ancestors for 70,000 years: that of scarcity. But everything comes at a cost: the globalisation of supply chains gives us consumer goods while taking away knowledge of their origin. We have gained in buying power what we have lost in buying knowledge. Is it any wonder that 16 million adults in the United States think that chocolate milk comes from brown cows?[66]

But not everyone is displeased as the West sleepwalks its way forward. For by organising the transfer of rare metals production, we have done much more than palm off the rare metals burden to the galley slaves of globalisation. We have entrusted a precious monopoly to potential rivals.

CHAPTER FOUR

The West under embargo

CELEBRATIONS WERE IN FULL SWING. THE WEST congratulated itself on its new ecological conversion, while deep in Jiangxi province the Chinese slaved in the gullies to crush the ore needed for this very transition. The lowliest work went to China, while in the West we focused on high-value-adding industries. We were winning at the game whose rules we ourselves had written and imposed.

That was before some geologists armed to the teeth with quantified reports turned up to break up our party. Indeed, this endangered species — for their numbers have dwindled in step with mining activities in the West — confronted us with an unpleasant and inconvenient reality. China, a preponderant producer of certain rare metals, could for the first time ever decide to cut off exports to the countries that most needed them.

Beijing: the new rare metals master

Every year, the United States Geological Survey, a government agency of the United States Department of the Interior, does an assessment of mineral resources for the preparation of its vital *Mineral Commodity Summaries*. In the report, analysts scrutinise over eighty raw materials considered critical to our modern economies.[1] Its 200 pages provide detailed statistics on resource availability, global stocks, and, most importantly, where in the world resources are mined.

The latter statistic is alarming: the USGS finds that Beijing produces 54 per cent of the antimony consumed worldwide, 58 per cent of the indium, nearly 66 per cent of the fluorspar, 67 per cent of the vanadium, and 73 per cent of the natural graphite.[2] This concurs with the European Commission's own list: China produces 67 per cent of the world's silicon, and 83 per cent of its germanium. It accounts for 86 per cent of tungsten production, and for between 85 and 100 per cent of rare-earth production.[3] European Commissioner for the Internal Market Thierry Breton concludes: '[F]or lithium, cobalt and graphite, Europe remains heavily dependent on supplies from third countries (China in the lead, not to name it), which can be as much as 100 per cent for refined lithium ... Above all, we need the capacity to separate, refine and recycle raw materials, which are also too often concentrated in China.'[4]

Following China's lead, myriad countries applying a specialist mining strategy have also acquired majority, if not monopolistic, positions. The Democratic Republic of the Congo produces 63 per cent of the world's cobalt; South Africa supplies 71 per cent of the world's platinum, 93 per cent of its iridium, 81 per cent of its rhodium, and 94 per cent of its ruthenium; and Brazil mines

92 per cent of the world's niobium. For its beryllium needs, Europe is dependent on the United States, which accounts for over 88 per cent of production. There are also countries whose share of global production is substantial enough to trigger temporary shortages and wild price swings. For instance, Russia alone controls 40 per cent of the world's palladium supply, and Turkey 48 per cent of the world's borate supply.[5]

Having the upper hand on rare metals is a question of survival for Beijing; the United States is not the only country very concerned about its supply security.[6] The reason is that China is not only the world's biggest minerals producer, it is also the world's biggest minerals consumer. To meet its needs, it guzzles more than 30 per cent of global industrial metals production.[7] (See Appendix 12 for China's share of global consumption of certain commodities.)

Chinese strategists are well versed in the challenges of mineral sovereignty. As a student in France, Deng Xiaoping worked in an iron foundry of Le Creusot.[8] As for his successors: 'The last six presidents and prime ministers — apart from the current prime minister [Li Keqiang, prime minister under Xi Jinping from 2013 to 2023], who read law — were all trained in engineering — electrical, hydroelectrical, geology — and in process chemistry,' a natural resources strategist told me in 2016.[9] Consequently, and with the support of a stable, authoritarian political system that values patient and consistent decision-making, Deng Xiaoping and his successors were able to lay the foundations of an ambitious policy to secure the nation's supplies.

The method was comparable to that of a steamroller: in the space of a few decades, China multiplied the number of mines in its territory, and launched works for a second 'silk road' on land

and sea as a raw materials supply route from Europe and Africa, while acquiring and merging companies in the commodities sector.[10] Global markets and the geopolitical status quo were steadily upended as Beijing's sphere of influence grew. More than simply taking its place on the global rare metals market, China underwent a complete transformation to become a maker of these markets.

It has become so powerful that any decision made in Beijing today has unavoidable global repercussions. One dip in local mining production, and the well-oiled wheel of supply and demand jams. A sudden jump in domestic demand is enough to trigger massive shortages. This is what happened with titanium — a mineral 50 per cent supplied by China globally. An unexpected rise in Chinese consumption between 2006 and 2008 increased titanium prices tenfold,[11] and put French aircraft manufacturer Dassault Aviation in a serious supply predicament.[12]

Chinese foreign policy's 'weapon of choice'

Beijing quickly realised the power it possessed from its stranglehold on rare metals. To put this into perspective, we can look at OPEC, the Organization of the Petroleum Exporting Countries. For decades, its thirteen members have been able to significantly influence the barrel price, yet they represent 'only' 35 per cent of global oil production.[13] China, on the other hand, has staked its claim on 100 per cent of global production of the coveted class of certain rare-earth metals. In the words of an Australian expert: 'It's OPEC on steroids.'[14] So what does a nation do when it realises just how powerful it is? Naturally, its intentions begin to take on a far more aggressive hue.

This is precisely what China is doing. The precepts of a hostile rare metals trade policy were reportedly outlined by Deng Xiaoping during a tour of the Bayan Obo rare-earths mine in the spring of 1992. 'The Middle East has oil; China has rare-earth metals,' he presaged. Chinese businessmen are known to smugly quote these words at raw-materials meetings and summits — words that say all that needs to be said.

At the beginning of the 2000s, close observers of the rare metals market noticed something not quite right. China's export quotas, set at 65,000 tonnes in 2005, began to drop a year later to just under 62,000 tonnes. By 2009, Beijing had further reduced this to 50,000 tonnes, and official figures for 2010 put exports at only 30,000 tonnes.[15] The same trend was observed for all the rare metals disproportionally produced by China. For instance, in August 2001, China enforced quotas on its molybdenum exports to the European Union. It then imposed a series of exorbitant export taxes between 2007 and 2008.[16] The WTO's analysis of the complaints against the Chinese was unequivocal: in the 2000s and 2010s, China was accused of engineering a policy of systematic restrictions on rare-mineral exports, ranging from fluorspar, coke, bauxite, magnesium, manganese, yellow phosphorous, silicon carbide, and zinc.[17]

In the 2000s, everyone from Jakarta, Los Angeles, Johannesburg to Stockholm began to feel the pinch from China. 'Every month, we worried about what quotas we could next expect,' says Jean-Yves Dumousseau (then sales director at Cytec), who worked in China at the time.[18] The hardest hit by the Chinese offensive was Japan — a major consumer of rare earths on account of its high-tech industries. Speaking anonymously, a Japanese diplomat I met in Tokyo told me: 'I attended many meetings between the Japanese minister of industry and Chinese government representatives. Whenever we raised the issue of rare earths, the Chinese made it plain that they could at any point turn off the tap [of exports].'

'There was no doubt that sooner or later a much bigger crisis would blow up in our faces,' an expert in the industry confirms. After all, isn't the twentieth century full of examples of states with a dominant position in strategic resources imposing embargos for commercial, diplomatic, or military gain?

One such example takes us back to the 1930s, when the United States imposed a helium embargo (of which it was the sole producer) on Germany for fear that the Nazis, already using the gas for their Zeppelin airships, would eventually put it towards aggressive ends.

Then, in 1973, OPEC declared an oil embargo against Israel and its allies in response to the Yom Kippur War, sparking the first oil crisis in history.

In 1979, US president Jimmy Carter halted the export of 17 million tonnes of grain to the Soviet Union in response to its invasion of Afghanistan.[19]

Such events have continued into the twenty-first century: in 2014, Russia cut off its gas exports to Poland and Ukraine multiple times due to diplomatic tensions.[20] And in 2022, in retaliation to European sanctions linked to the conflict in Ukraine, Moscow drastically tightened and even stopped its gas exports to numerous European countries.[21]

After gas, oil and then grain were wielded as weapons, it was inevitable that China would weaponise its metals. And so, in September 2010, it launched an embargo on rare earths that defies belief.

It was also to be the first embargo of the energy and digital transition.

Trade manoeuvres with global repercussions

Behind the massive trade fallout is a longstanding dispute between Japan and China over the Senkaku (or Diaoyu) Islands — an archipelago of five small islands and three rocks in the East China Sea, north-east of Taiwan. While it might not sound like much, the area conceals vast amounts of oil and gas, which is why the two Asian powers have coveted the islands since the end of the nineteenth century.

Japan took the Senkaku Islands from China after the First Sino-Japanese War in 1895. They were then placed under US control at the end of the Second World War, before being handed back to Japan in 1972. This has not stopped China from asserting its sovereignty over the territory — and incurring the wrath of the Japanese. So, when a Chinese trawler dared to cast its nets near the coasts of the islands on 7 September 2010, the Japanese coastguard saw this as a provocation and gave chase. The scene that followed — filmed and available online — is captivating: refusing to comply, the Chinese captain steered his trawler into a collision with the Japanese patrol boat.[22] His detention by the coastguard caused an uproar in China. Quick to tug at any nationalistic thread, the Chinese media had no difficulty in sustaining the nation's indignation over the incident.

Was it any surprise, therefore, that two weeks later, on 22 September, all deliveries of Chinese rare-earth metals to Japan were halted, and without any official declaration of an embargo? 'The incident with the trawler sharpened our nationalistic

instincts,' Chen Zhanheng, vice secretary-general of the China
Rare Earth Industry, told me in Beijing. 'Several Chinese
companies took it upon themselves to suspend their deliveries
to Japan.' Astonishingly, Chen Zhanheng made no bones about
the fact that the Chinese authorities, not wanting to upset the
WTO, denied that any official embargo had been imposed.

A year after the events, Japanese businesses still did not
believe a word of this. So I headed 2,000 kilometres away from
the Forbidden City to Japan. Pulling out of Tokyo station in the
Shinkansen high-speed train, I slowly skirted round Mount Fuji
and its conical outline jutting into the autumn sky. Four hours
later, I began to make out the tentacles of Osaka, the country's
third-biggest city, extending out along the Pacific coast. I had
come to talk to Kunihiro Fujita, a rare-earth metals importer, at
his factory warehouse to hear his version of events.

'China has always used its natural resources as political
leverage,' he said. Wearing a dark suit and a hard hat, Fujita
showed me his stockpile of yttrium — a rare earth used for
precision electronics. But in September 2010, his orders were
suddenly no longer honoured by his Chinese suppliers. 'The
Japanese industry was in a panic,' Tokyo University professor
Toru Okabe told me. Rare-earth metals are the 'vitamins' of its
high-tech industry. They are so important that 'even a cleaning
lady knows what they're about'.[23] What had started as a banal
incident at sea turned into a catastrophe for Japan.

Soon the crisis took on an international dimension. In the
days that followed, many European and US rare metals importers
also began to worry about the sharp decline in Chinese exports.
The Western media, up till then largely unfamiliar with these
tiny metals, sunk their teeth in as well, headlining 'international

tensions', a 'showdown' between China and Japan, and a 'war' over the acquisition of this 'crucial category of minerals' that are 'more precious than gold' and used in cutting-edge industries.[24] The EU trade commissioner's spokesperson stressed that rare earths were a 'major concern' for the European Commission, and appealed to China to 'allow the markets to operate without hindrance'.[25] The US secretary of state, Hillary Clinton, spoke on the matter at a press conference in Hawaii, and announced an imminent visit to China to resolve the crisis.[26] Some weeks later, Jean-Louis Borloo, the French environment minister under the Nicolas Sarkozy presidency, published a ministerial order to create the Committee on Strategic Metals to address the risks of shortages of metals critical to French industry.[27] And on the steps of the White House, President Obama announced America's submission of a trade complaint to the WTO against China.

The rare metals war had begun.

Back in Osaka, Kunihiro Fujita tells me how he continued to suffer the effects of the 'informal' embargo for six months after the arrest of the captain of the trawler. That is, until it occurred to China that it, too, might find itself short of high-tech consumer goods 'made in Japan' that could no longer be exported due to a lack of resources![28] But in the meantime, the rare metals market was gripped by panic as it was hit with the reality of the supply shortage. This, and the hand-wringing over Beijing's moves, together with the speculative behaviour of certain Chinese traders, caused prices to skyrocket, with a host of other rare metals following suit.[29] Brokers, traders, and importers in all four corners of the globe spent most of their time trying to extract any semblance of a promise to deliver the materials or, failing that, passing on the consequences of these unprecedented disruptions to unsuspecting customers.[30] 'It was utter chaos!' says Fujita. 'The natural course of supply and demand no longer had its place.'[31]

Journey to the queen of platinum

The new 'metals risk' is not linked to China's export policy alone. A surge of nationalism over mining resources is sweeping across Asia, Africa, and Latin America, and is increasingly weakening Western positions.

As I discovered in 2009, nowhere is the groundswell more apparent than in some of the most hidden parts of southern Africa. Johannesburg, the economic capital of South Africa, was my next port of call. From there, I drove three hours along vast expanses of savannah to Phokeng, in the North West Province. At first glance, everything in this isolated town of some 15,000 inhabitants in the middle of the bushveld appeared normal. Yet it was anything but. Gigantic totems and multicoloured flags bearing the official emblem of the crocodile reached into the sky, while a group of men wearing the blue uniforms of the Royal Bafokeng Reaction Police Force patrolled the pristine streets.

There was no customs house or border post. Nor any sign indicating that, just a few kilometres earlier, I had crossed the invisible border of the Royal Bafokeng Kingdom — a territory covering 1,400 square kilometres, almost the land area of Greater London.

While the kingdom is an integral part of the 'rainbow nation', it has also established its own governance, administration, clan structure, budget, and customary law system. The reason for this, and also why I had come, lies several hundred metres underground: the most extraordinary deposits of platinum group metals — ruthenium, rhodium, iridium, platinum, and more. These rare and precious metals have multiple applications, ranging from jewellery and laboratory equipment to catalysts for cars.

From Phokeng, we cut a few kilometres north through the veld to reach the Rasimone mines, its ground pocked and swollen by the extraction of rocks from below the surface. Standing before us, in the middle of the bush against a rolling landscape, were the towering iron behemoths of the platinum refineries. Railway tracks interlaced the facilities, their carts of waste rock hauled by trundling locomotives overhead.

'When I started here, we only mined on one level. Now we're ten levels down!' exclaimed Dirk Swanepoel, a white South African miner working for Anglo Platinum. A stone's throw from his office, a dozen or so miners wearing overalls and hard hats climbed off a conveyer belt coming out of a chasm made in the rock. They had been using pneumatic rock drills to bore holes that would subsequently be blown up using explosives. The ore brought to the surface is crushed and reduced to particles of rock containing platinoids. They are then plunged into water containing special reagents, decanted, dried, melted, and purified to obtain platinum. 'Every month, we extract 200,000 tonnes of rock,' Swanepoel told us. 'Each tonne contains about four to seven grams of platinum.'

The Bafokengs happen to be sitting on the biggest platinum deposit in the world — a treasure that the 'People of the Dew' were unaware of when they settled in the fertile lands in the fifteenth century. In 1870, King Kgosi Mokgatle acquired the first 900 hectares of the current territory, using the fortune amassed by his subjects in the diamond mines of neighbouring Kimberley. The precious metals were discovered in 1924.

After the birth of democratic South Africa, the Bafokengs — wronged by the segregationist regulations of apartheid, which prohibited them from owning their land — entered a long legal battle with Impala Platinum (Implats), which had been pocketing the proceeds from mining on the territory. The Bafokengs emerged triumphant, receiving up to 22 per cent in royalties, and acquiring a stake of nearly 13 per cent in the company. Since then, they have adopted a financial-diversification strategy, of which mines represent a shrinking portion.[32] It was the first time a South African ethnic community had come up trumps against a mining company.

Leruo Molotlegi, who has been the *kgosi* (king) since 2000, and the thirty-sixth monarch of the dynasty, is overseen by the queen mother, who has a symbolic role. Semane Molotlegi was one of the last queens of the 'dark continent'. She was virtually invisible in the media, and it was almost unheard of for a foreign journalist to meet her. I was fortunate to be granted an audience with her.

I dressed up to the nines to meet with the exquisite woman in her fifties, who was draped in elegant, colourful traditional dress, her voice the essence of poise. She is known respectfully as *Mmemogolo* (meaning 'grandmother' in Setswana). We spoke about her travels around the world to spotlight the success of her people.[33]

The Bafokeng contradict the curse of raw materials, by which Western societies ship in, use up the local resources, and ship out. Instead, the Bafokeng became the richest tribe on the continent, and even plan to diversify their sources of income.[34] The strategy they implemented is textbook-worthy: by standing up against mining companies, the people of platinum asserted the supremacy of the producer over the buyer, and of the owner — sovereign over its resources — over customers around the world. This case is also unprecedented in that it was a mere 'tribe', rather than a state, that took on a multinational.

While the Bafokeng baulk at the idea of resource nationalism, this is nevertheless a case study of a rebalancing, if not a reversal, of the traditional power relationship. The mining companies, often acting on behalf of Western consumers and used to imposing their terms, understood they would increasingly be outplayed and would therefore need to adapt. The landmark example of the Bafokeng captured the attention of international organisations, and the nation was paid a visit by the World Bank, the World Economic Forum, United Nations agencies, and US academics.[35]

The resurgence of mining nationalism

This is far from being an isolated case. Today, more and more states are refusing access by foreign industrial companies to promising mining areas. In 2013, the Mongolian government halted the operations of Rio Tinto in the Oyu Tolgoï copper mine in the Gobi Desert.[36] More recently, in 2022, Chile's new government withdrew the lithium mining permits that had been granted to two private companies under the previous government.[37] Other countries banned foreign companies from buying local mining companies: in 2010, the Saskatchewan province in Canada stymied an attempt by Anglo-Australian company BHP

Billiton to buy out the Canadian group PotashCorp, the world's leading potash producer.[38] At the end of 2022, citing reasons of national security, Canada went so far as to ask three Chinese companies to divest their shares in several major Canadian mining companies (Teck Resources, Ivanhoe Mines, and First Quantum Minerals) involved in lithium exploration ... before backpedalling four months later.[39]

States are also investing in traditionally private mining groups. In 2023, Saudi Arabia, through its state-owned mining company Ma'aden, acquired a 9.9 per cent stake in the US mining exploration group Ivanhoe Electric, a move that will enable it to speed up exploration of its own subsoil, whose mineral wealth is estimated at $1.3 trillion.[40] And what is Qatar trying to achieve by multiplying its stakes via the state-owned Qatar Mining Company or the Qatar Investment Authority, the emirate's sovereign investment fund? Doha has already invested in mining groups such as the Swiss multinational Xstrata.[41] It has also landed exploration permits for deposits in Mali, Burkina Faso, and, more recently, in Kazakhstan.[42] The minute Gulf state seems to be on a quest for metals for which it has no apparent need.[43]

Lastly, and most importantly, restrictions on the free trade of metals are on the rise. Indonesia, like China, offers another eye-opening example. In 2009, the undeniable mining powerhouse declared a series of embargos on the export of more or less all the raw minerals produced by the archipelago. In 2014, it upped the ante by including nickel, tin, bauxite, chromium, gold, and silver. 'We had to ensure the sovereignty of our raw materials,' explained a senior government official I met in Jakarta. 'Any political action concerning our mining wealth must be determined and executed by our government and not by foreign states.'[44]

There are many more examples, as observed by the Organisation for Economic Co-operation and Development (OECD). Numerous studies devoted to trade in raw materials list the restrictions placed on commodity exports around the world. From 2009 to 2020, thousands of international trade violations were documented. The OECD produced a fascinating chart that shows how measures have increased since 1961: they were at relatively low levels until 2005, but then the curve went up sharply, and has not come down since.[45]

We see these restrictions in almost all the minerals and metals included in the OECD inventory. In 2021, Argentina imposed barriers on the export of its mineral resources by applying taxes on borates, copper, iron, lithium, and strontium, just as South Africa has done with diamonds; India, with bauxite, iron, and titanium; Kazakhstan, with aluminium; and Russia, with tungsten. Certain countries go further, such as Mongolia, which has banned all copper exports; Tajikistan, all aluminium exports; the Philippines, all gold exports; and Zimbabwe, all chromium exports.[46] What is driving this widespread trend?

Most mineral-producing countries have developed to become emerging countries. Governments now have to consider their burgeoning middle class and the more prosperous and resource-hungry consumers they represent. The growing sentiment is that locally mined resources should be used for domestic consumption rather than to satisfy the appetites of buyer countries. There is also a growing ecological conscience and activism — once the domain of the West — that oppose local mining plans. This has resulted in states having to tighten social and environmental regulations, which makes it take longer to bring a mine into operation.

More broadly speaking, a culture of resistance is taking root from Jakarta to Ulaanbaatar, from Buenos Aires to Pretoria, as a newer and savvier middle class is wiser to what they see as a sell-off of their resources. They are spearheaded by political leaders who observe an economic rebalancing between developed countries, often bogged down in stagnation, and dynamic emerging countries craving wealth. Protectionist measures are therefore demonstrations of power in a 'de-Westernising' world.

This is not new: from the 1960s, the wave of independence in the third world came with claims for sovereignty over resources.[47] In Africa, in 1958, Ghana nationalised the Ashanti goldmines that were formerly under British control. The Democratic Republic of the Congo followed suit in 1965, as did Tanzania in 1967, Zambia in 1970, and Zimbabwe in the 1980s. Then came the liberal wave of open trade. But it took the Chinese policy of slapping quotas on rare metals exports to reignite — and amplify — resource sovereignty across five continents. 'China galvanised the nationalism of resources,' says an American expert, 'not only on its own territory, but all over the world.'[48]

In the years following China's embargo in 2010, it was no longer a question of *if* new trade crises would occur, but rather *when* they would occur. Between taking sips of tea in the bar of a luxury hotel in the Shanghai suburb of Xuhui, Vivian Wu, a highly authoritative voice in the rare metals industry, told me in 2016 that such a scenario had every chance of coming true: 'Rather than embargos, I prefer to talk about action measures against Japan and other countries. These actions form part of a strategy led by the Chinese government to restore our image.

And they may be applied again in the future, be it for rare earths or other metals.'[49]

She could not have put it better: in 2019, in the midst of the trade war when President Donald Trump banned Chinese giant Huawei from the US telecoms market, Xi Jinping, accompanied by Liu He, his vice-premier in charge of economic affairs, had his photo taken at a JL Mag Rare-Earth factory, a producer of rare earths in the southern province of Jiangxi.[50] Not a word was uttered by the Chinese head of state, but the message couldn't have been clearer: were trade tensions to escalate between the two superpowers, Beijing could retaliate by again cutting off rare-earth exports to its rival. China's official state news agency immediately confirmed as much: '[w]aging a trade war against China, the United States risks losing the supply of materials that are vital to sustaining its technological strength'.[51]

These threats have unnerved Western countries, and not for the first time, as the reliance of their strategic industries on raw materials produced primarily by China is further exposed. As are the Middle Kingdom's formidable ambitions in this latest episode in the rare metals war: an economic conflict in which Beijing harnesses its leadership in the production of minerals such as graphite, gallium, indium, tungsten, antimony, and, of course, rare earths, to limit their export and, as we shall soon discover, to produce its own technologies packed with these strategic raw materials, and to challenge the West's technological superiority.[52]

Metals of influence, metals of crisis
The specific qualities of the rare metals markets could make matters worse:

- As we have seen, they are incredibly narrow markets: output is derisory compared to that of the markets for major metals such as iron, copper, aluminium, and lead. For instance, the global production of rare earths barely makes up 0.015 per cent of steel production.[33]

- They are highly confidential, and involve very few buyers and sellers. Any move on their part therefore is all the more likely to upset the balance of supply and demand. Accordingly, a default by just one supplier can throw consumers into a tailspin, just as a new technology requiring rare metals can cause a sudden shortage in supplies.

- They are opaque markets in which business discretion and an absence of formalities are the rule. Other than a handful of metals listed on the London Metal Exchange (LME), there are no official reference prices, and everything is traded over the counter. Buyers often need to consult specialised journals or connect to Weibo, the Chinese microblogging website, where brokers and traders drip-feed information about their latest transaction amounts.

- Moreover, these markets are strategic for mining countries. China has proven very reluctant to provide certain production data that are considered state secrets.[34] Hidden stockpiles and geostrategic and diplomatic factors render the market even more obscure, even for leading analysts.

- And, finally, the sudden increase in private investors, in it for their own interests, has also impeded the interplay of supply and demand. For all resources, a natural resources specialist explained, in 2012 these third-party players traded 'sixty times more "raw material" stocks than ten years ago', which has contributed to price instability.[55] Traditionally, the major base metals were the subject of market speculation, but increasingly the rare metals market is, too.[56] Speculators include 'hedge funds [such as the Tudor Fund in the US], asset managers [such as PGGM Investments in the Netherlands], pension funds [such as Pacific Investment Management Company (PIMCO) in the US], as well as the finance departments of US universities [such as Harvard and Princeton]', according to a banking analyst.

Any major position taken in a metal results in wild speculation. One of many examples:[57] the purchase in 2017 of 17 per cent of global cobalt production — that is, several thousand tonnes — by investors betting on a shortage of cobalt brought on by a sharp price increase.[58] And in 2022, uncertainty over nickel production, of which Russia is a major producer, caused nickel prices on the LME to jump from an average of $15,000 in 2020 to $20,000 in January 2022, and to $48,000 two months later. Prices have cooled down significantly since, but remain between $22,000 and $30,000 a tonne.[59]

Making any kind of forecast about these ultra-sensitive markets is therefore near impossible. 'The rare-earths market is neither stable nor predictable,' Vivian Wu emphasises. Gone are the days when, even without regular supplies of rare metals, governments and businesses could at least count on constant market prices to sketch out a strategy. As a specialist from a European geological institution says: 'Rare metals are crisis metals.'

CHAPTER FIVE

High-tech hold-up

MONOPOLISING MINING WAS BEIJING'S FIRST VICTORY. BUT it wasn't long before China began to turn its sights to downstream high-tech industries that use rare earths.

Battle of the super magnets

This process started with magnets, the metals of which I first encountered in July 2011 when visiting Peter Dent, the director of magnet manufacturer Electron Energy Corporation, at his company's headquarters in the small town of Landisville, Pennsylvania.

During my visit, Peter guided me past the warehouses to a storage area bathed in pallid neon lighting. 'There you have it: rare earths!' he exclaimed as he turned on the timer light. Dotted across the polished concrete floor were greenish barrels containing lumps of grey and slightly corroded material: samarium samples, gadolinium pellets, and other metals with perfectly unpronounceable names. After years of tracking these metals without ever laying my eyes on a single one, I felt like a buccaneer before Blackbeard's bounty. I had every intention of taking my time to explore this treasure trove.

More interesting still were the adjoining workshops. Dent continued the tour: 'This is the machine room.' In its din, workers spent all day machining small, round electromagnets containing rare metals. 'This is where we shape and size our magnets. We produce hundreds of thousands of them every year,' Dent said. After a well-oiled manufacturing process, the magnets were lined up on trays like bread loaves fresh out of the oven, and carefully loaded into carts.[1]

Until the mid-1970s, there were only a handful of industrial applications for rare earths and other rare metals — for their luminescent properties, they were used in lighters and camping lanterns.[2] Their scope of application first grew with the advent of colour television screens.[3] The real game-changer, however, was rare-earth magnets: developed in 1983, these pure marvels of technology have become indispensable in all products equipped with electric motors, reputed to be pollution-free.[4]

We know that an electrical charge coming into contact with the magnetic field of a magnet generates a force that creates movement. Traditional magnets made from the iron derivative ferrite needed to be massive to generate a magnetic field powerful enough for more sophisticated applications. 'Remember how we used to walk around with mobile phones the size of bricks?' jokes an expert. Just one of the many shortcomings of an oversized magnet.

Meanwhile, in the mobility industry, the race for lighter and more efficient energy had begun. Engines needed to be as light and compact as possible — reducing the calibre of an engine meant reducing the proportions and weight of the object in which it was used. Such progress would without a doubt generate massive energy savings.[5]

This very progress was made possible by rare-earth magnets, revolutionising modern electronics. In fact, without even realising it, you have probably already come into contact with these super magnets, especially if you happen to have a magnetic knife-holder on your kitchen wall. Have you ever wondered how a single magnet can defy gravity to suspend a 20-centimetre steel blade in the air? Certainly not with the help of ferrite. 'At equal magnetic strength, a rare-earth magnet is 100 times smaller than a ferrite magnet,' an industry expert explained. 'This is about miniaturisation; rare earths make objects smaller.'[6] It's also about making electric engines powerful enough to challenge the dominance of internal-combustion engines. It gave the energy transition and digitalisation a formidable kickstart.

This is also precisely where the trouble started.

Rewind to the 1980s. Rare-earth magnets are all the rage, and have colonised manufacturing sectors the world over — giving Japan, and its electronics company Hitachi, which holds the patent, the unassailable lead in the industry. So much so, in fact, that 'the Japanese banned the export of this technology to China', Chen Zhanheng told me.[7]

Beijing wasn't put off in the slightest by the embargo. It decided that, in addition to making off with virtually all rare-earth resources, it would start to take control of the miracle technology behind the end products. In this way, explained Chen Zhanheng, 'China's own industries could benefit from the added value of rare-earth minerals.' And by whatever means.

In the 1980s, magnet manufacturers were mostly situated in Japan, from where they supplied the bulk of global demand. But they could no longer resist the siren song of their Chinese counterparts, who offered to take over the menial work of machining the least-sophisticated magnets. 'The Chinese said, "Come to Canton! Relocate your low-value rare-earth applications, and we'll take care of your low-tech!" an Australian consultant told me.[8]

The Japanese may have had the technology, but the Chinese had the allure of cheaper production that would allow Japanese businesses to expand their profit margins. The Japanese didn't hesitate for long. Boasting full employment and a strong currency at the time, the island nation considered it a sound decision. The history books will look back at this and say that Japan, the second-strongest global power at the time, knowingly exported to its competitor the technologies it lacked.

As related in Chapter Three, French chemical company Rhône-Poulenc was enticed by the prospect of low-cost rare-earths transformation, and also moved part of its refining operations to China. It did this by setting up joint ventures with Chinese partners from the 1990s. This had trade unions up in arms, and it would have taken an insurrection and a fight to the bitter end to keep their jobs in France. But Rhône-Poulenc had other priorities. Its pharmaceutical branch was about to be privatised to become Aventis. Therefore, 'the strategic and geopolitical importance of this small [chemical] business was completely lost on them', recalled a former employee.[9]

Jean-Paul Tognet, who worked at Rhône-Poulenc at the time and made several trips to China at the end of the 1970s to look into future partnerships, knew what was brewing: 'The technical assistance our partners wanted from the West was a one-way street: they expected everything to be handed over to them. They found it completely normal to be helped ... without giving anything in return.'[10] He assured me that Rhône-Poulenc did not divulge any industrial secrets. But by abandoning their refining sector to the Chinese, only to then become their most loyal customers, the West — and France in particular — handed China the market on a silver platter.

In the 1990s, low-end refineries began to mushroom — first in the Baotou region and then in the rest of China. 'Across the country, rare earths had become the goose that laid the golden egg. Money flowed, and factory bosses drove around in Lincolns!' Jean-Paul Tognet recounted.[11] Essentially, we had created the ecosystem our opponents needed to reproduce Western knowledge, make massive profits, invest in their own R&D activities, and thus break into the downstream industry at full force. Jean-Yves Dumousseau, a chemist who worked in China at the time, shot straight from the hip: 'Rhône-Poulenc slipped China's foot into the stirrup.'[12]

But where was the harm? After all, Rhône-Poulenc believed it had a twenty-year head start on China. By discontinuing the more trivial aspects of its refining business, it could move up the value chain to more sophisticated intermediary products (particularly in luminescence). But in 1987, one of its engineers noticed the staggering progress being made by Chinese refiners, and put his foot in it: 'I told them we were only three years ahead of the Chinese at best. It caused an uproar!'[13] Only his timeline was off. 'By 2001, Chinese refiners were all on a technological par with us,' Dumousseau said.[14] 'We might have underestimated this competitive risk for too long,' Tognet ventured. 'The Chinese wanted to move up the value chain, and we couldn't stop them! We cared more about rock-bottom production costs, especially with customers breathing down our necks ... And of course, today nothing has changed.'[15]

In 2023, only a few rare earths are still transformed in La Rochelle. The separation workshops have closed their doors, and the site's operations are a shambles. The group's revenues have decreased, and headcount, which was at 630 employees in 1985, has halved.[16] Can Rhône-Poulenc (today Solvay) at least take pride in new downstream prospects that sustain jobs in France? 'Even those businesses were transferred to our facility in China ... It all comes down to cost!' Tognet said. Perhaps we can take comfort in the fact that Solvay still has interests in joint ventures with its Chinese partners. This may be, but, as Tognet bitterly put it, 'Solvay now sees itself as a Chinese company as well!'[17]

China's awakening set in motion unavoidable economic disruptions. But did we really need to give Beijing that extra leg-up? Our poor assessment of China's competitive streak, combined with our outright quest for profit, without a doubt precipitated the transfer of labour, work units, and, most importantly, technologies to China.

A chronicle of impending deindustrialisation

Underlying our apparent blindness are numerous instances of wishful thinking. The West has long held to the illusion of eternal scientific progress — a philosophy that has irrigated several economic sectors since the 1980s. We believed that by abandoning our heavy industries, we could focus our efforts on high-value-adding manufacturing sectors while maintaining healthy profit margins. Some believed that emerging countries would continue to be the factories of the world, churning out jeans and toys, while we would hold sway over the more lucrative segments. 'I think most people continued to believe that only blue-collar jobs would be affected by the shifts [brought on by Chinese competition],' a trade unionist from the US metal industry told me. 'We didn't realise that we were going to lose more than coffee-cup production, and that the more skilled jobs would be hit much harder economically.'[18]

Added to this is the pipedream of manufacturing fading into the background in favour of a service economy. The focus should be on knowledge and the immense added value it generates. This belief, echoing the utopia of a dematerialised world discussed in Chapter Two, was heavily subscribed to by the business world at the turn of the twenty-first century. Like Alcatel-Lucent's CEO Serge Tchuruk, many US and European business leaders could not resist the allure of 'factoryless companies'. Grey matter was more valued and therefore given more support, to the detriment of the tool that is the lowly factory. This line of thinking led to the separation of business from factory, as the former opted for outsourcing. In France, this trend was largely driven by the love lost between citizens and industry. 'At the start of my career,' Régis Poisson, a former engineer at Rhône-Poulenc, told me, 'a factory worker could become famous for something he designed. Then came the rejection of enterprise and the deterioration of the company image. Today, the working class no longer likes factories, which symbolise exclusion.'[19]

Thus, the West and China took this path hand in hand. But from the 2000s, the Chinese began to use far less conventional methods: rare metals quotas that rattled the magnet manufacturers that hadn't relocated their factories (to guard their industrial secrets). As their rare-earths provisions ran dry, they were faced with a difficult choice with equally painful outcomes: keep their industrial activities at home at the risk of having to slow down production for lack of raw materials, or relocate to China and have access to abundant supplies.[20] For Japan, the dilemma was short-lived, a London analyst explained to me: 'They were starved of raw materials. So they took their technologies to China.'[21]

Beijing's treatment of those determined to hold their ground was cruel: price distortion. This was much to the fury of Sherrod Brown, a US senator from Ohio, who, in a fiery speech in 2011, said, 'China is organising artificial shortages and export quotas to raise prices internationally while keeping them low domestically. How can we compete when the Chinese are so brazenly cheating?'[22] The same unflinching observation was made in 2022 by Marco Rubio, the Republican senator from the state of Florida. A few months earlier, Rubio had introduced a bill to reshore rare earth processing operations to American soil.[23] Before an audience of US financiers, he said, 'Rare earth minerals, which really aren't that rare, will never be mined or exploited in the United States, because the Chinese will never allow it. If it's a pure market decision, no matter what we price it at, they will undercut us. We have to confront that reality.'[24]

This tactic was equally unjust for the vast majority of magnet manufacturers outside China. At the end of the 1990s, Japan, the US, and Europe made up 90 per cent of the magnet market, but now China controls 92 per cent of global output![25] Using blackmail in the form of a technology-versus-resources trade-off, China expanded its monopoly in mineral production to include mineral transformation. It therefore dominated not one but *two* stages of the industrial value chain. This is confirmed by my Chinese rare-earths expert Vivian Wu: 'I'd even go as far as saying that in the near future China will have an entirely integrated rare-earths industry, from one end of the value chain to the other.'

In fact, this wish has already partially come true. And nowhere more so than in the city of Baotou, in Inner Mongolia.

Journey to the 'Silicon Valley of rare earths'

In Chapter One we saw behind the scenes of the world's rare-earths capital, Baotou. We saw its toxic lakes and its cancer villages, whose inhabitants are dying a slow death. Let us now examine its glittering exterior.

It's a Saturday, and by day the city's elegant glass towers cut a sleek figure against the mineral desert. By night, when darkness falls over the plain, 'the little Dubai of the steppes' adorns itself in blinking lights and glowing lanterns, repelling the cold obscurity of the surrounding landscape. Jianshe Road, Baotou's main stretch, fills up with people who have come to admire the shop window displays and to peruse the fragrant food stalls lining the pedestrian alleys. There is a sense of triumph and conquest here; I can see it in the smug smiles of passers-by, and even in the buildings still wrapped in plastic sheeting — all symbols of a victorious metropolis convinced of its fantastic destiny.

It's an idyllic setting for a delegation of eighty international businessmen invited to Baotou to attend an international conference on rare earths in October 2011. Their hosts are clearly out to impress, having put up their guests in a luxury hotel with suites overlooking the verdant parks of the city centre. Even the cloudless skies seemed to have been ordered for the occasion.

In the conference room, Sun Yong Ge, a high-ranking official in charge of the Baotou Economic Development Zone, continues the charm offensive: 'Baotou is the rare-earths capital! We welcome technology industries – we can supply them with virtually all the minerals they need.'

The city has made technology the cornerstone of its development, its centrifugal force being its close proximity to the rare-earth deposits where China takes its fill. 'We want to be more than just suppliers of raw materials; we want to supply more elaborate products,' Yong Ge says. Western businesses that, like the colonisers before them, sought only to mine resources to generate added value back home are no longer welcome in Baotou. However, the apparatchik says, 'we are very open to transformation companies wanting to move their technologies to China'.

Whether out of interest or for lack of choice, numerous foreign manufacturers have already converged on the mineral refineries in the 120-kilometre free zone on the city's outskirts. An international presence is reflected in the numbers: according to Sun Yong Ge, every year Baotou produced 30,000 tonnes of rare-earth magnets. Today that number stands at 50,000 tonnes, and is expected to reach 100,000 tonnes by 2025.[26] As it happens, the delegation was invited to visit the plants. Journalists, however, were kindly requested to stay behind. So I entrusted a small camera to Jean-Yves Dumousseau, who came back with images of a power display. 'This is where they make the magnets you have in your iPhones and iPads,' he commented as we watched the video. 'The Chinese paid an absolute fortune for these technologies copied from European know-how.'[27]

By orchestrating the transfer of magnet factories, the Chinese accelerated the migration of the entire downstream industry — the businesses that use magnets — to the Baotou free zone. 'Now they've moved onto producing electric cars, phosphors, and wind turbine components. The entire value chain has moved!' confirmed our witness. Sun Yong Ge added to the list: '10,000 tonnes of polishing material, 1,000 tonnes of catalytic converters, and 300 tonnes of luminescent materials.'

This makes Baotou much more than just another mining area. In fact, the Chinese give it the well-earned title of the 'Silicon Valley of rare earths': in 2023, the city had over 8,500 companies, nearly 100 of which specialise in rare earths.[28] Also coming out of the industrial zone is top-of-the-range equipment, and the hundreds of thousands of people working there generate $7.5 billion in annual revenues.[29]

When China decided to assume the burden of the 'oil of the twenty-first century' three decades ago, it did not focus on manufacturers on the decline nor turn its back on the high-tech sector. It chose to cross swords with the West over resources so that, one generation later, it could set its sights on the high-end digital and green-tech industries. Thus, rare-metal restrictions did more than serve China's sporadic embargos. The second stage of its offensive is far more ambitious: China is erecting a completely independent and integrated industry, starting with the foul mines in which begrimed labourers toil, to state-of-the-art factories employing high-flying engineers. And it's perfectly legitimate. After all, the Chinese policy of moving up the value chain is not dissimilar from the viticulture strategy of winemakers in the Napa Valley in California, or the Barossa Valley in South Australia. As one Australian expert put it, 'The French don't sell grapes, do they? They sell wine. The Chinese feel like rare earths are to them what vineyards are to the French.'[30]

This strategy of moving up the ladder is not limited to rare earths. As early as the 1990s, a wave of concern rippled through the fabric of German small-to-medium-sized enterprises (known as the Mittelstand) specialising in the manufacture of machine tools. (Machine tools are used for factory work automation, from basic milling machines to ultra-connected machining centres.) The Mittelstand acted pre-emptively by gradually replacing humans with robots, allowing it to remain competitive to the extent that German industry still represents almost a quarter of the country's GDP.[31]

Except that industrial robots require terrific amounts of tungsten. China has always produced this rare metal in abundance, but there are other tungsten mines around the world, ensuring supply diversity for manufacturers. During the 1990s, the Chinese machined their own cutting tools — 'Some hammers, a few drills ... really crumby tools,' said an Australian consultant.[32] But they wanted to move up the value chain in this area as well. 'They drove down tungsten prices [from 1985 to 2004],[33] hoping that Westerners concerned about getting their raw materials at the best price would buy exclusively from the Chinese, and that competing mines would shut down.'[34]

We can guess what could have happened next: the Middle Kingdom — now the hegemonic power in tungsten production — would have used the same blackmail tactic to force the Germans to move their factories as close as possible to the raw materials. The Chinese would have crushed any German lead in the cutting-tools industry, and would then have made off with the machine-tools segment — a pillar of the Mittelstand. It would have been the hold-up of the century!

But the Germans saw the Chinese coming, and aligned instead with other tungsten producers (Russia, Austria, and Portugal, among others). 'They preferred paying more for their resources to sustain the alternative mines and not depend on the Chinese,' the Australian consultant told me.[35]

No matter. China applied the same modus operandi to the graphite market, where it is also a notoriously preponderant player. The mineral graphite, the purest form of carbon, is used to produce graphene —a nanomaterial a million times finer than a single strand of hair, yet twice as strong as steel. Its discovery by physicists Andre Geim and Kostya Novoselov earned them the Nobel Prize in Physics in 2010.[36] Recognising the massive markets that graphite was creating, China 'pursued a strategy similar to that of developing the downstream value chain', explained Vivian Wu. China's commercial policy already includes export duties and quotas to protect its domestic market.[37] Recognising the danger, the US filed a new complaint at the WTO in 2016 against Beijing's practices that 'disadvantage US producers by raising the prices of these raw materials for manufacturers outside of China, while lowering the prices paid by China's manufacturers that use these same raw materials'.[38]

By now a pattern is emerging, and it is being applied to molybdenum and germanium, a Chinese journalist I met in Beijing told me.[39] Between 2009 and 2020, export restrictions on lithium, cobalt, and manganese increased ninefold.[40] 'They're using the same industrial policy for iron, aluminium, cement, and even petrochemical products,' warned a German industrialist.[41] In China, there is even talk of applying this policy to composite materials — new materials resulting from alloys of several rare minerals. What if it developed a miracle composite material that the rest of the world could not do without? China certainly wouldn't sell it any less sparingly than its other resources. In addition to keeping a blacklist of critical minerals, the European Union should list all critical alloys whose supply could be threatened.[42]

The West is starting to put words to what has happened with China: whoever has the minerals owns the industry.[43] Our reliance on China — previously limited to raw materials — now includes the technologies of the energy and digital transition that rely on these raw materials. 'Is this a non-military conflict? The answer is most assuredly yes!' said one US rare-earths expert.[44] Are we on the winning or losing side? The answer of a specialist from the French mining industry is cutting: 'We're not even putting up a fight!'[45]

Unsurprisingly, other mining states around the world are following China's lead. A case in point lies in the middle of the island of Java, in the Indonesian capital of Jakarta.

Indonesia's new 'non-aligned movement'

If there is one capital in the world that could be given the title of Hell City in the twenty-first century, it would be Jakarta. Located in the immense archipelago of Indonesia in South-East Asia, and nicknamed 'The Big Durian' in reference to the foul-smelling fruit eaten by Indonesians, Jakarta is a city you endure rather than visit. And of all the senses vexed by this intolerable megalopolis of 30 million people, touch was without a doubt the most affected: the muggy heat saturating the air; the unrelenting rainfall harassing this city of glass and concrete; and the close shaves of my motorcycle taxi with the hordes of vehicles tearing down the main roads.

Because it was impossible to distinguish north from south, or to familiarise myself with any particular landmark — be it a tower, intersection, or major road — I mentally choreographed my own: cable-constricted bamboo trees; an armada of motorcycles straddling the torrents of water gushing out from the drains; the pungent smell of garlic invading the palm tree–lined alleys; the cadence of a train trundling overhead; the vestige of a forest forgotten between two buildings. And on it went.

A few days later, while coming in to land at Bangka, 400 kilometres further north, I had my first glimpse of the purpose of my trip that winter in 2014. From the plane, I saw thousands of craters — as if a meteor shower had rained down on this island the size of the Greater Paris area. They are in fact tin mines, below whose surface thousands of miners play their part in a thriving black market. There are also mines offshore, recognisable by the thousands of small wooden barges on the surface. From these crude vessels, young men plunge 20 metres deep with nothing more than a breathing tube connected to an air compressor. They scrape the seabed, sending the sand to the surface, with the help of a makeshift vacuuming device. From their barges, raw material is separated from minerals using a small sorter.

Bangka is the world's biggest producer of tin — a grey-silver metal essential to green technology and modern electronics, such as solar panels, electric batteries, mobile phones, and digital screens.[46] Every year, over 300,000 tonnes of tin are mined around the world. Indonesia represented almost a quarter of global production in 2021, making it the second-biggest exporter of the high-tech mineral, which is nevertheless not considered rare.[47] The archipelago recognised the value of this outstanding mineral: from 2003, as a spokesperson for one of Indonesia's biggest mining houses, PT Timah, explained: 'Tin became the first mineral to be used in an embargo.'[48]

It would be the first of a very long series of embargos. From 2014, all of Indonesia's mineral resources — from sand to nickel, and diamonds to gold — were no longer exported in raw form. As explained by Indonesian authorities, 'The minerals we don't sell now will be sold tomorrow as finished products.' As in China, this policy was a powerful way to generate wealth. By some calculations, preserving the added value in this way quadrupled profits on iron, increased profits on tin and copper sevenfold, bauxite profits by a factor of as much as eighteen, and nickel profits by as much as twenty.

The Indonesians did far more than replicate the Chinese model: they innovated by laying the foundations of financial nationalism. In 2013, Jakarta established the Indonesia Commodity and Derivatives Exchange (ICDX) to fix the tin price, shunning the 'diktat' of the world's biggest metals markets, the LME, in doing so. 'Our goal is to control and stabilise the market price,' explained Megain Widjaja, the young director of ICDX, who felt that the tin price was regularly manipulated. As such, all tin exported by Indonesia would first have to go through the Jakarta stock exchange.

The impact of this policy is still being debated. According to Widjaja, annual tin prices in 2016 fluctuated by only 8 per cent, compared to 20 per cent and 30 per cent previously. But according to a London analyst, the prices set by the LME still serve as a benchmark, and he didn't believe that it would change anytime soon.[49] He did concede, however, that 'Indonesia applied a very original idea'. Indeed, to support its industrial policy, the nation needed to build road networks, electricity grids, ports, train stations, and airports. It would take strong and stable mineral prices to pay off and maintain these investments.

So Indonesia warded off the invisible hand of the market by using the controls of the stock market. This inspired other Asian markets. In 2015, the Shanghai Futures Exchange included tin among the metals for trade on its futures market,[50] and announced in 2018 that it would soon include copper.[51] Malaysia followed suit in 2016.[52] Other trading floors also introduced trading platforms for copper, nickel, and zinc.[53]

But Indonesia's nationalisation of its mining resources at the time did not meet with the same success as that of its Chinese counterpart, for it failed to put the necessary means behind its policy. The colossal investments needed to build the downstream industry were delayed, and the country's trade balance began to teeter as budget shortfalls accumulated. In 2017, the country had no choice but to loosen its strategy by allowing several minerals to be exported again.[54] The fall in commodity prices was also largely responsible for this setback, as this nationalistic approach had, for the most part, been implemented during a global 'commodities super cycle' — fifteen years of plenty that began in 2000, during which market prices skyrocketed. It was a boon for buyer countries, but also put seller countries in a position of force that aroused their nationalist instincts.

But the super cycle ended in 2014. The consumer-producer

trade power returned to an even keel, and producer countries began to think twice before investing higher up the value chain. However, this setback did little to dampen Jakarta's spirits. At the turn of 2020, it renewed its nationalist policy by banning exports of nickel, of which it is the world's leading producer, with the stated aim of forcing buyers to set up refineries in the archipelago.[55] Result: between 2020 and 2023, Indonesia signed a dozen commercial contracts worth $15 billion with foreign groups. Some say that the old world is resilient, and that it will have the last word. That is, unless emerging countries begin to consume as conspicuously as OECD countries do, giving rise to a new world that will eclipse three centuries of Western order.[56] Yet how do we make ourselves heard by billions of individuals who dream of eating meat at every meal, of drinking champagne, and of travelling to Europe and the US with their families to take a selfie in front of the Eiffel Tower or the Empire State Building?

For emerging countries, rare metals are more than ever a means to a better way of life. It is an unstoppable reality that is only growing: in 1998, after years of struggle, the Kanak separatists of New Caledonia obtained the majority holding in a refinery in the Koniambo massif — the world's biggest nickel reserve. Local populations therefore benefit from local processing of the mineral — synonymous with added value. A similar trend is emerging in Cambodia, Laos, and the Philippines. According to Mostafa Terrab, the CEO of the OCP Group, Africa is also joining the trend: 'There is a growing phosphate-processing industry in Africa to manufacture fertiliser for Africans. It is likely to spread to other sectors. Africa has no choice but to industrialise.'[57]

This intuition is consistent with the 2050 African Mining Vision advocated at the twelfth summit of the African Union in 2009.[58] The aim is to make mines a driver of inclusive growth by capturing a bigger share of their added value. To date, only 15 per cent of African mining production is believed to remain on the continent. We are far from an optimal result, for over 80 per cent of mining production from Sub-Saharan Africa is exported from the continent.[59] But given Africa's growing contribution to global GDP, change is inevitable. More importantly, the stakes are no longer purely industrial or political. The fair distribution of minerals — a shared global asset — has become a *moral* imperative, and a cause in which international institutions, such as the World Bank, are now united.[60]

CHAPTER SIX

The day China overtook the West

THESE CONTENTIOUS METALS HAVE LAID BARE A BATTLE OF invention. Today, nations are competing for the brightest minds and most innovative start-ups, and want to claim ownership of the most outstanding patents to epitomise their culture and genius. New technologies promote a certain economic and societal model or lens through which to view the world. China knows this. Through its industrial rare metals strategy, it is betting heavily on scientific development, encouraging the creativity of its people, and ultimately cultivating an alternative civilisation model to the tenets dictated by the West.

Chinese recipes for state-directed high-tech

The theoretical foundations of China's intellectual emulation were laid in 1976, when Deng Xiaoping broke ranks with Mao's agricultural ambitions to proclaim that, from then on, 'production power lay in the sciences'.[1] Subsequent leaders have perpetuated and deepened this conviction: in 2006, President Hu Jintao stated that 'science and technology' form 'the central thread in the development strategy of China'.[2]

These declared aspirations translated into action in 2010 with the twelfth Five-Year Plan — a road map tracing the key economic strategies for the 2011–2015 period — which identified as priorities seven advanced industries and just as many new-technology horizons;[3] five years later, innovation and technological progress formed the centrepiece of the thirteenth Five-Year Plan (2016–2020);[4] and in 2021 the fourteenth Five-Year Plan's strategic focus areas included new materials, renewable energy, robotics, aerospace, satellite communications, and electric mobility.[5] Thus, concepts that were never central to China's history had become mantras of the state.[6]

To consolidate this vision, Beijing drew on the terrific competitive advantages of its economy: cheap labour from the inner provinces; cheap capital, attributable to China's policy of devaluing the yuan; and the sheer size of the Chinese domestic market that allows significant economies of scale.[7] Beijing also accelerated competing companies' relocation of their production facilities, using partnership as its weapon. Under so-called joint ventures, foreign businesses gained access to the abovementioned competitive advantages in exchange for their technological know-how and, therefore, their patents. Beijing called this absorption and internalising of foreign technologies 'indigenous innovation'.[8]

The strategy's foundations were set out in an industrial policy document published by the Chinese government in 2006.[9] An American consultant based in Beijing ironically described the document as 'full of ... good intentions and gilded rhetoric about international cooperation and friendship'.[10] The reality is that China's definition of indigenous innovation is reworking and adjusting imported technologies to develop its own technologies. 'The plan is considered by many international technology companies as a blueprint for technology theft on a scale the world has never seen,' a US report published in 2010 asserted. It continued: 'With these indigenous innovation industrial policies, it is very clear that China has switched from defense to offense.'[11] Successive US Democratic and Republican administrations — from that of Obama, to Trump, to Biden — have tried to counter this strategy. Following on from his predecessor's trade war with China, Joe Biden enacted a bill on 5 January 2023 to protect American intellectual property against Chinese offensives.[12]

The Chinese applied this very tactic to rare-earth magnets: they enticed — or forced — foreign businesses onto its territory under the guise of joint ventures, and then launched a process of 'co-innovation' or 're-innovation'. This is how China purloined the technologies of Japanese and US super-magnet manufacturers.

Having reaped the benefits of the invention of others, Beijing built an ecosystem of endogenous creation to 'move from factory to laboratory', starting with a variety of research programs that began in the early 1980s.[13] One of the most emblematic of these, the '863 Program',[14] was launched to make China a leader in seven leading industries, many of which are considered 'green'.[15] The 2015 'Made in China 2025' plan saw the creation of some forty industrial innovation hubs across the country.

In 2021, the Chinese state spent over $500 billion on research — less than the US ($645 billion), but more than the European Union ($360 billion).[16]

But China has many weaknesses: relative to its population size, it has far fewer researchers than France or the UK; there remain colossal challenges to education; while rural China — a massive part of the country — is sidelined from this momentum. And does the state that has so far managed to combine state interventionism and entrepreneurial freedom run the risk of impeding innovation? More than anywhere else, the success of the innovation ecosystem hinges on administrative efficiency. But achieving this will require the government to commit to painful structural reforms with unclear outcomes. Moreover, the inertia of state-owned enterprises — a formidable part of the energy, telecommunications, and finance sectors — is no longer viable ... yet often their leaders are top government officials. How will the government reform these conglomerates without creating tensions and deadlocks within the party?[17]

Lastly, some of China's characteristics do little to aid its cause. While an interventionist regime may have allowed a strategic state to flourish, it leaves no room for any deviation. How can an administration that employs two million government agents and, in 2020, spent $6.6 billion dollars to restrict online freedom of expression encourage creativity?[18] A government that stymies the freedom to criticise — and therefore to think differently — nurtures a potent culture of copying, and turns the lack of inventiveness into a building block. Another excellent example of this is the small portion of China's R&D budget allocated to fundamental research: just 5 per cent compared with 7 per cent in the US and 24 per cent in France.[19] 'The Chinese have the technology, but they are stuck in organisational and ideological thinking that is still poorly adapted to the modernity of the twenty-first century,' concluded an interpreter and former Western diplomat posted to Beijing.[20]

Be that as it may, the Chinese authorities are doomed to succeed. Furthermore, innovation, together with the logic of moving upmarket, presents the communist regime with two real, shorter-term challenges:

- The first stems from its strong desire for technological independence, fuelled by a litany of past humiliations. The first dates back to the fallout from the Sino-Soviet split at the end of the 1950s: on the back of diplomatic tensions, in the summer of 1960 the USSR withdrew its technical assistance, on which many heavy Chinese industries depended for their long-term survival.[21] Then, in 1989, in retaliation for the repression of the student movement on Tiananmen Square, the US imposed an embargo on the sale of weapons to China. Beijing learned the painful lesson that it could not rely on anyone's strengths but its own. Since then, an obsession with self-sufficiency has pervaded the Chinese psyche — so much so that China aimed to have decreased its reliance on foreign technologies to 30 per cent by 2020, from 60 per cent in 2006.[22]

- The second challenge is the survival of the Communist Party of China, which has concluded a tacit contract with one-fifth of humanity: the acceptance of an authoritarian regime in exchange for rising standards of living — a pact that a severe economic slowdown would render null and void. To hold up its end of the bargain, the state has to integrate up to 15 million workers who enter the urban job market every year — a target that will almost certainly have been met in 2020, 2021, and 2022, according to Chinese government sources, thanks to the creation of 12 million new urban jobs.[23] It's an unrealistic feat if China focuses all its efforts on mining alone — indeed, there are more jobs and even greater growth potential downstream of the value chain. Rare metals therefore constitute one of the keys to the resilience of an authoritarian regime that, on top of hardening its policy of technological surveillance, must constantly innovate to avoid going the way of many imperial dynasties before it.

Astounding technological progress

These factors combined help to explain why the Communist Party of China has so far been so incredibly successful. 'I was in China in the early 2000s. Back then, we spoke about textiles, toys, assemblies for electronic products. Quite frankly, no one could have imagined what happened next,' admits a European journalist based in Beijing.[24] China's astounding progress in the electronics, aerospace, transport, biology, machine tools, and information technology sectors caught everyone off guard — including the upper realms of the Communist Party.[25] In aerospace, China has already landed a robot on the moon with the first core module of its own space station, Tiangong;[26] it also plans to land an astronaut on the moon by 2030.[27] In 2022 alone, it launched some sixty-four orbital missions, dethroning Russia as the US's main competitor in the new space race.[28] Beijing wants to move beyond the demand side of new technologies by trading its status as a skills consumer for that of a skills supplier.[29] This policy is reflected by the staggering number of patent applications out of China in 2022: 750,000 out of a total of 1.5 million patents worldwide — more than any other country in the world.[30] (See Appendix 15 for the number of patents per country per year.)

And while we cry over spilt milk, the Middle Kingdom is picking up the pace: it wants to explore the still-unknown properties of rare earths to develop the applications of the future. Some of its university research programs are advanced enough to both astonish and alarm a researcher at the US Department of Defense: 'Losing our supply chain was tragic enough. But now China is busy getting a ten-year head start on us. We could easily find ourselves without the intellectual property rights of the applications of the future that matter the most.' China is openly pursuing this move: 'We want to use these metals to become the world leader in technology,' Vivian Wu said.

On 29 September 2010, Kathleen Dahlkemper, a member of the US Congress, had already said as much in the House of Representatives during China's embargo of rare earths: 'The Chinese have taken control of the rare-earths market, and we, the United States, are being overtaken.' Her words were crucial: Dahlkemper did not say that the US's technological advance could be reduced or diminished. She did not say that China was catching up with the world's superpower or that the Chinese would put more and more obstacles in its way. She said that we, the West, were about to be overtaken, as is already the case in a growing number of industrial segments.

In fact, Beijing has already designed the J-20, an ultra-sophisticated stealth fighter jet;[31] dominated the race, from June 2013 to November 2017, for the world's most powerful supercomputer;[32] and put into orbit, in 2016, the first quantum communications satellite with reputedly impregnable encryption technology.[33] On a broader scale, a recent report by the Australian Institute for Strategic Policy, a think tank funded by the Australian and US governments, reports that China has taken the lead in thirty-seven of the forty-four cutting-edge technology sectors analysed, including fifth- and sixth-generation communication technologies (5G and 6G), electric batteries, and nuclear energy.[34]

Above all, China has taken the lead in an impressive range of green tech. It has put its image as a polluting and polluted country behind it to stand out as the world leader in green-energy generation: it is the number-one manufacturer of photovoltaic systems (over 80 per cent in 2022);[35] the global hydropower giant;[36] the biggest investor in onshore and offshore wind power;[37] and is the world's largest market for new-energy vehicles, representing over half of electric vehicles on the world's roads. (See Appendix 16 for the geographic distribution of the EV value chain.)[38] Between 2018 and 2023, China will have manufactured an estimated 13 million electric vehicles, while Germany, the second-biggest EV manufacturer, will have manufactured 4.4 million, and the US 4.1 million.[39] The undisputed winner of this reversal of power with the West is the BYD [Build Your Dreams] Company — China's flagbearer, with almost a million EVs sold in 2022 alone.[40]

Announcing federal measures in 2022 to support the US critical metals sector, President Joe Biden gave a perfect reading of China's strategy of moving downstream to the point of the technological frontier in the industrial segments of the energy transition: 'China controls most of the global market in these minerals. We can't build a future that's made in America if we ourselves are dependent on China for the materials that power the products of today and tomorrow.'[41] Is it still possible to turn the situation around? By 2030, China is expected to manufacture more than double the amount of electric vehicle batteries than all other countries combined.[42] With a monopoly in the production of rare metals and in the green technologies that depend on them, China intends to remain the world's biggest green-tech producing country. It wants to siphon green jobs away from Europe, Japan, and the US.

Economically speaking, it wants to stand out as the outright winner of the energy and digital transition.

This ambitious ecological shift will also soothe growing tensions in China over environmental matters that have become unacceptable for the population.[43] Indeed, anti-pollution protests now number between 30,000 and 50,000 every year. Whether protesting against a petrochemical complex project, such as in the city of Kunming, Yunnan, or opposing the construction of a waste incinerator, such as in Hangzhou, Zhejiang, a middle-class movement aligned with the global NIMBY (Not In My Back Yard) trend is openly rejecting the Chinese growth model. There are close to 8,000 environmental non-profits working to harness, coordinate and unify this groundswell.

'The old development model cannot last,' environmentalist Ma Jun said to me. 'We cannot continue consuming the way we do. Change is needed.' This green movement is working to modernise the country's growth drivers by giving preference to 'lighter' services and technology, such as digital technologies that have a smaller environmental impact. This will give China a greener image on the global stage, and even place Beijing as the diplomatic leader of the energy transition, thus filling the void left by the withdrawal of the United States from the Paris Accords.

The diminished West
The Chinese strategy of moving up the rare metals value chain was executed at the expense of the industrial vitality of Europe and the US. It implicitly revealed the vulnerability of the Western economic model, despite its establishment as a benchmark at the end of the Second World War.

A German academic put numbers on this reality: looking only at rare earths since 1965, the capture of the oxide market (the powders refined by Rhône-Poulenc) has led to the transfer of US$4 billion of wealth to China.[44] Further up the value chain in the magnet and battery markets, this figure increases tenfold to over US$40 billion. It makes sense: there is greater value for China in capturing the downstream chain.

An Australian researcher applied this reasoning to even more advanced manufacturing sectors, such as the components industry — the assembled parts integrated into a consumer good (such as printed circuit boards, sensors, amplifiers, diodes, LEDs, thermostats, and switches).[45] Following the same logic of the higher capital gain inherent in downstream industries, the transfer of wealth from the rest of the world to China is estimated to have increased tenfold to US$400 billion. The researcher then considered equipment manufacturers — the industries that produce parts that are even more sophisticated than components (in the car sector, this includes dashboards and built-in cameras; in IT, computer hard drives; in aerospace, engines, and software for commercial aircraft). The amount of transferred wealth is again multiplied by ten to reach US$4,000 billion — twice France's GDP, or the equivalent of Germany's GDP.

The annual estimated rare-earths market worth is a derisory $10 billion — 220 times smaller than the oil market.[46] But because these small metals are found in just about everything we use, this microscopic industry takes on gigantic proportions. We should also consider that neither of the two studies accounts for the loss of mining industries and end-product manufacturers (such as wind turbines, electric vehicles, and solar panels); the loss of tax revenue for governments; the impact on trade balances; or similar consequences of China moving up the value chain of other rare metals.

And what of the millions of jobs created on one side at the expense of millions lost on the other? In the US, one need only drive an hour out of Chicago along Lake Michigan to the neighbouring state of Indiana to see the disastrous effects of the Chinese battering ram on the US metallurgy industry. In the city of Gary, I met Jim Robinson, a United Steelworkers trade unionist, in his office. His cheerless face told a story of the demise of what was a thriving steel industry until the 1980s. 'The region was so industrialised that it was called the American

Ruhr!' he recalled, in reference to Europe's first industrial basin in western Germany. 'It was in its heyday. No one could have imagined what we were in for.' For its residents, Gary is 'a wreck': entire neighbourhoods have been abandoned; houses — their doors hanging off their hinges — are going for US$50. Every day, men and women flee this industrial ghost town ravaged by unemployment, despair, and insecurity.

Three-quarters of magnet manufacturers in the US are gone. The sector that once numbered 6,000 professionals in its ranks in the US now employs no more than 500.[47] Further downstream, Japanese car manufacturer Toyota and its German competitor BMW have also moved some of their operations to China.[48] Tesla's plant in Shanghai is its biggest gigafactory, with an annual output of 75,000 vehicles.[49] Japanese conglomerate Sumitomo, the German chemical company BASF, and its companion in misfortune, W.R. Grace in the US, have taken the same route. 'Naturally, they were drawn in by China's cheap labour costs. But the biggest push factor for moving their factories was access to rare earths. Millions of jobs were siphoned,' Australian academic Dudley Kingsnorth explained.[50] By betting on renewables, Beijing precipitated the downfall of a fossil-fuel-based industrial order in which the West excelled in favour of a new energy system in which the West is already lagging behind. In a way, one can understand Donald Trump's refusal to commit the US to the energy transition: he preferred to preserve an oil-energy model that allowed the US to dominate the twentieth century[51] than commit to going fully electric — a move he knows could be detrimental to American industry.[52]

France was not spared either. Arnaud Montebourg, the then minister of industry, reported as much in his op-ed piece published in a major French newspaper: 'In 2001, a small village in Haute-Garonne, Marignac [in south-western France], painfully witnessed the loss of France's only magnesium factory to Chinese competition. Many years later, with its monopoly on the magnesium market, China increased its production prices to a level that would have returned the French facility to profitability. In the interim, France had lost hundreds of jobs ... in car and aircraft lightweighting.'[33]

A lack of interest in the metal-refinement industry also led, in 2013, to the liquidation of the French operations of Comptoir Lyon-Alemand, Louyot et Cie — a company specialised in processing precious metals. 'It was the only French rare metals company,' said a former employee who worked there as a non-ferrous metals buyer and precious-metals trader. 'Nearly 4,000 jobs were cut in 2002' in France and globally.[34] The jobs at stake in all the high-tech sectors were highly specialised, requiring advanced expertise. 'They were the "green jobs" that would create a modern and developed economy,' an American trade unionist added.[35] With them went know-how — some of it centuries old — with ramifications for such sectors as defence, electronics, automotive, and, of course, the energies of the future. These human and social dramas add to an already bleak picture: 900,000 industrial jobs lost in France over the last fifteen years, or a 25 per cent drop in the workforce. Over the same period, the secondary sector's contribution to France's GDP slid four points.[36] The figures are not much better in Europe and the US, where the industry made up 22 per cent and 18 per cent of GDP respectively in 2018, down 11 per cent and 12 per cent since the turn of the century.[37]

Deindustrialisation in the United States and Europe

hammered the post-war social contract, stirring up deep social turmoil and creating a hotbed for a franchise of populist parties. Donald Trump succeeded in reaching the White House because he could count on the voters in the deindustrialised states of the Rust Belt. In these swing states, where votes can tip the result of a national election, the Republican candidate vigorously denounced the anti-competitive practices of the Chinese and offshoring, and emphasised the need to protect the US from the industrial war spearheaded by Beijing. The strategy paid off, and Trump won the popular and Electoral College vote in virtually all these states, wiping out the comfortable popular-vote majority enjoyed by Hillary Clinton at the national level.

The West would have felt the pernicious effects of de-industrialisation with or without rare metals. But this economic, social, and political crisis was amplified by China's monopoly of the resources destined to replace fossil fuels, and by its formidable strategy of absorbing the green industries reliant on rare metals. Moreover, the European model has proven 'powerless to implement a policy that preserves its economic, technological, and social assets', writes a Western expert, adding: 'The survival ... of European democracy ... may be the last hurdle in the way of the scarcely begun emergence of Chinese industry.'[38]

When two world views collide

Meanwhile, China's success has allowed it to foster a model of government that values a long game, unlike the short-sighted decisions of the West that have wrecked most of its industrial policies. This 'authoritarian capitalism ... provides encouragement to other autocratic states', suggests an Indian academic.[59] It has proved it can produce solid growth while guaranteeing political stability, thereby giving substance to the 'Beijing consensus' — the idea that the Chinese development model can serve as a benchmark for other emerging countries.[60]

This consensus opposes another, that of Washington, which has prevailed since the end of the Cold War, and by virtue of which economic growth and democratic progress are necessarily correlated. The war for rare metals — and green jobs — has therefore lifted the lid on a new ideological conflict that pits China's principles of political organisation against those of the West.

'A clash of civilisations is a very Western way of seeing things!' says Zhao Tingyang. Zhao is a philosopher who became famous in China by popularising *Tianxia*, a concept inspired by the teachings of Confucius, which promotes the search for harmony in international relations.[61] He had agreed to talk to me about the future of China's relationship with the West. He said that the centrifugal force of globalisation 'pushes us to become more and more interdependent, making conflict — military or economic — an undesirable option for everyone'. In this world at peace, the philosopher predicted, *Tianxia* will reign; it will be a new era of power led by the globalisation of communication and transport, which will unite a cosmopolitan ruling elite and share Western and Chinese values. 'I believe this system will create peaceful interactions and better common understanding,' he said.

At a large table laden with fine food, I met with a Chinese rare metals heavyweight who concurred. 'The world to come is more open and cooperative,' he said in between chopstick manoeuvres.[62] A corollary to this feel-good statement is the

serenity of China's 'panda diplomacy', which involves gifting
specimens of these giant folivores to countries with which it
wishes to forge diplomatic ties. You would have to be heartless
not to fawn over online videos of China's favourite furry mammal
rolling about and eating bamboo. With the panda, China seeks
to convey the idea of a peaceful emergence, at all costs.

It's tempting to believe in this, as it is to dream of an
ethnically diverse world at peace. Except when the game is given
away by one detail: by acquiring a strategic rare-earths company
in Indiana from under America's nose, China has whipped the
covers off its impressive military program.

CHAPTER SEVEN

The race for precision-guided missiles

HOLLYWOOD HAS TAKEN A FERVENT INTEREST IN RARE earths. And rightly so. Rare resources as indispensable as oil, the threat from China, and the survival of high-tech industries ... It's blockbuster material. This certainly wasn't lost on the screenwriters of *House of Cards*, the hit television series set in the 2010s about Frank Underwood (played by Kevin Spacey), an American politician who ruthlessly climbs his way to power in the White House. Rare metals are woven into the storyline in the second season, which bears a few similarities with reality.

In one episode, China uses its 95 per cent monopoly of the global production of samarium-149 — portrayed as a very rare metal that is indispensable to the operation of American nuclear reactors — to make the US pay a hefty price for the resource.[1] This could make the price of electricity prohibitive for consumers, and Washington is plunged into political disarray. On Frank Underwood's suggestion, the US decides to bypass the Chinese monopoly by buying samarium through a third party. This forces the Chinese, says the protagonist, 'to lower [their prices] to keep a direct flow with us. We stockpile samarium for defence purposes ... and quietly sell off what we don't need to our friends ...'

For the Pentagon, there is nothing new about stockpiling rare metals needed for the country's defence as vital components of its war arsenal: tanks, destroyers, radars, smart bombs, antipersonnel mines, night vision equipment, sonars, or the new laser weapons already tested in the Persian Gulf.[2] (See Appendix 17 for a list of rare metals used in a fighter aircraft.)

More pressing still is the growing strategic importance of these resources as the threat of cyber warfare escalates. Indeed, twenty-first-century warfare is increasingly taking place on multiple fronts. As we see in the 'hybrid' war led by Russia against Ukraine, and against Western states in general, belligerents strike not only on land, but in the air, in space, in cyberspace, and through media channels by seeking to wipe out the enemy's communication channels, control images, rewrite history, and manipulate opinions.[3] Warfare has moved from the ground to the stratospheres of electronic, media, and virtual wars, with rare metal–hungry servers, drones, radar aircraft, constellations of satellites, and space launchers as arsenal.[4] The further away we move from the battlefield, the deeper we need to dig below it.

In terms of physical volumes, armed forces' rare metals requirements remain very low. 'In 2023, only 1 per cent of France's metal requirements got to defence,' says a French military expert, speaking anonymously. And according to an American expert in metals for technology, the US defence industry imports a total of 200 tonnes of magnets every year — just 0.25 per cent of global production.[5] Yet the world's powerhouse, whose military under

President Joe Biden is expected to monopolise $842 billion of the US budget in 2024, would be cut down to size should these strategic components not make it to the arms factories.[6]

The subject gained in importance following Russia's invasion of Ukraine on 24 February 2022. In September 2022, 'the Pentagon became very concerned about the availability of certain critical materials used in the manufacture of the weapons they supply to Ukraine', points out the same French expert. What's more, Russia is one of the main producers of several strategic metals, including antimony, which is essential to the manufacture of American munitions — a situation that has led the Department of Defense to support the reopening of a mine for this specific metal on American soil.[7] Moscow also supplies 40 per cent of the world market for aircraft-grade titanium. While European sanctions have conveniently eliminated imports of this essential resource for aircraft engines and landing gear, the European Union nevertheless recognises the pressing need to diversify supplies (by turning to South Africa in particular).

And the list of Western armies' dependence on external suppliers is even longer. 'It includes gallium [essential for the manufacture of night vision systems] and tungsten [used to manufacture shells, grenades, and shotguns],' says our expert. 'China produces 60 per cent of this metal, Russia 20 per cent ... and we Europeans have no reserves. We're completely exposed, and there may well be a crisis in this metal in the future'.

Pooch parlours and precision-guided missiles

Back to rare earths. For decades, the Pentagon sourced its magnets from manufacturers in the US. One of the most strategic of those was a company called Magnequench. By specialist accounts, it was the best rare-earth magnets producer in the world: its factories were the pinnacle of the production line for Abrams battle tanks and for Boeing's JDAM smart bombs, which were used in the Afghanistan and Iraq wars.

The highly prized supplier of the US Army ran its successful outfit out of Valparaiso — a charmless town of 32,000 people in the state of Indiana, a two-hour drive from Chicago. Former Magnequench employee Terry Luna agreed to escort me to what was once the manufacturing area. Although somewhat the worse for wear, the buildings, tall and rigid in the humid summer heat, were unchanged. But their occupants were different. 'I started at Magnequench at the front desk,' explained the full-figured woman in a nasal voice. 'But if you look at the sign, it says, "Coco's Canine Cabana": a doggy day-care centre.'

We went in and were given a warm welcome by the centre's manager, a young blonde woman wearing a T-shirt and ripped jeans. We walked across a hall, since repurposed for holding birthday parties for pooches, and through a grooming parlour complete with doggy shampoo and clippers — 'made in China' (of course). We then reached the central storage area. 'Here they kept magnets for bombs, remote-controlled missiles, and a whole lot of war weapons,' explained Terry, barely holding back her tears. 'And they sold the lot! It all went to the dogs!' In 2006, Magnequench closed its strategic plant in Valparaiso to open a new one — in Tianjin, 130 kilometres south-east of Beijing.[8] 'The Chinese learned our industrial secrets,' added Terry. Two hundred and twenty-five employees were laid off. It's a tragic snapshot of an America whose most qualified workers abandoned their manufacturing tools to make way for pets.

Magnequench's fate paints a picture of Beijing's new military ambitions while it imposes its status as an industrial power. China's defence budget is now second to that of the US, spending US$292 billion in 2022 compared to US$105 billion (nearly one-third lower) in 2010.[9] China's goal? To go as far as dethroning the US by 2049 — the year that the People's Republic of China celebrates the centenary of its founding.

Since the 1980s, the Chinese army has developed in three ways. First, it changed its doctrine when Deng Xiaoping abandoned the dogma of 'the people's war', according to which battles took place in the heart of the Chinese territory. This was replaced with 'the people's war under modern conditions', whereby armed forces fight at the country's borders and beyond. The next change was organisational, as the army moved from the idea of 'the power of many' to a smaller, specialist army. The last change was technological, after the Gulf War left China facing the reality that it needed to catch up with the US.[10] To achieve this third transformation, Beijing undoubtedly gave much reflection to history's golden rule about natural resources: peace and metals rarely make good bedfellows.

Six thousand years ago, when humans began to melt copper, they were able to put aside their stone tools for sharper and more solid implements. At best, humans could perfect their hunting techniques with this breakthrough until 2,000 years later, when the Sumerians discovered bronze — a robust alloy of copper and tin. This time, empires and civilisations were emboldened to turn out swords, knives, and axes, raise armies, and enter what would be the very first arms race in history.[11]

Around the twelfth century BC, in the south of modern-day Turkey, the Hittites melted an even lighter and more widely available metal — iron — to forge weapons that were more powerful and easier to wield. This, say some historians, led ultimately to the European conquest of the Americas from the fifteenth century.[12]

Then came steel, which in 1914 tipped Europe into an industrial war. The iron and carbon alloy was used to make shell casings, the first modern fragmentation grenades, hardier helmets for soldiers, and armoured tanks — all of which contributed to the bloodbath that was the First World War.

Every time a people, civilisation, or state masters a new metal, it leads to exponential technical and military progress — and deadlier conflicts. Now it is rare metals, and in particular rare earths, that are changing the face of modern warfare. China knows that the master of their production and application will undeniably have the upper hand strategically and militarily. Targeting Magnequench and acquiring its patents and, therefore, its secrets was a perfectly logical move.

But the Magnequench affair gave rise to grave national-security concerns, as exposed to the public in 2015 by the CBS (Columbia Broadcasting System) television program *60 Minutes*. All it took was the factory's removal from US soil to put the world's military powerhouse at the mercy of Beijing for the supply of some of the most strategic components of its military technology. How could America have put itself in such a precarious position?

Magnequench in the radar of the 'Red Princelings'

'I spent twenty-one years in the office of the Secretary of Defense, working on tech transfer ... and in those twenty-one years, the

Magnequench case ranks among the top five of the thousands of cases that I've seen.' Peter Leitner was a senior official at the US Department of Defense at the time of the Magnequench affair. His job was to review all technology exports that might have a negative impact on America's military sovereignty, and to stop the export if required. The buyout of Magnequench was in fact already underway in the 1990s, during the presidency of Bill Clinton — years before the factory was moved. Magnequench's parent company, General Motors, agreed to sell the magnet manufacturer to the Chinese in exchange for approval to build a vehicle plant in Shanghai.[13]

During that time, Leitner and his colleagues were caught up in another matter: the sale by Kentucky-based US company Crucible Materials of its rare-earth magnet-production operations to YBM Magnex International, a company listed on the Alberta stock exchange. On paper, YBM seemed like a perfectly above-board business, with a head office and warehouse in Philadelphia. Upon closer inspection, however, it turned out to be a shell company owned by the 'Red Mafia' — a Russian criminal network thriving in the void created by the fall of the Soviet Union.

YBM was, in fact, hand in glove with a mob of shady businessmen, including a Ukrainian national by the name of Semion Mogilevich, whom the Federal Bureau of Investigation (FBI) would later describe as 'a global con artist and ruthless criminal ... involved in weapons trafficking, contract murders, extortion, drug trafficking, and prostitution on an international scale'.[14] Under the guise of selling magnets, the company busied itself with laundering the proceeds of criminal activities perpetrated in Russia and in numerous other states of the former Eastern bloc.

Crucible Materials nevertheless sold its magnet operations to YBM on 22 August 1997, from which time a part of America's production of the magnets needed for its defence was at the mercy of organised crime. By putting itself in the running for an industry as strategic as this one, did YBM make use of a geopolitical agenda dictated at a higher level to serve its own interests? Russia's state of decline at that time hardly put it in a position to oversee such an operation, and the motives of the key players in the affair are vague to this day.

The stakeholders of the Magnequench buyout proved to be just as dubious. Meet Archibald Cox Jr, the man who bought Magnequench. The son of a special prosecutor during the Watergate scandal and the chairman of the private equity firm Sextant Group, he was also a seasoned trader who was drawn in by the alluring operation.[15] 'He was a very slick figure,' recalled Leitner. 'He wasn't shy about charming those in charge at the Department of Defense who had concerns about the deal. He'd say, "Don't call me Cox, call me Archie!" They were seduced by his charm, and were at that point predisposed to approve anything he wanted. It was just incredible.'

But Peter Leitner and his colleagues were quick off the mark; they knew Archibald Cox and the Sextant Group were the screen between Washington and Beijing. Through a subtle misdirection of funds into tax havens in the Caribbean, dubious Chinese businessmen had been brought into the deal. 'One of them was the "Red Princeling" [the heir to one of the top officials of the Chinese Communist Party]. It was none other than Deng Xiaoping's son-in-law.'

It is public knowledge that Deng Xiaoping handed the reins of the China Nonferrous Mining Group Corp (CNMC) — a formidable state-owned mining company — to his son-in-law Wu Jianchang to ensure the longevity of the strategic nonferrous metal sector.[16] It was the very same Wu Jianchang who, through CNMC's New York branch, was directly involved in the sale of Magnequench.

Peter Leinter later discovered that not one, but *two* of Deng Xiaoping's sons-in-law were involved. Zhang Hong, married to Xiaoping's daughter Deng Nan, was chairman of Beijing San Huan New Materials High Tech Inc. — Magnequench's final buyer. This wasn't just any buyout of an American company. 'All these things were riddled with ambiguity,' said the former civil servant, who reported the problem to his superiors. 'We pointed this out repeatedly. We went to our superiors within the defense department, we argued ad infinitum, and eventually we were ignored.'

The apathy of the Democratic administration is even more surprising given that China was in the process of executing its 'Sixteen-Character Policy', which it had outlined as early as 1978.[17] Enacted by Deng Xiaoping, the strategy included a program to acquire 'dual-use' technologies to ultimately strengthen China's army. Magnequench was an ideal choice, as its magnets were used for the vehicles of General Motors and for the US Army.

The premise of the Sixteen-Character Policy was pragmatic: given the difficulty in procuring war technologies due to the US arms embargo, China would buy foreign companies whose know-how in civil applications could be repurposed for more hostile ends. In the years that followed, this strategy would lead to an extraordinary proliferation of Chinese espionage against the US. According to a former US counterintelligence agent, 'China's intelligence services are among the most aggressive [in the world] at spying on the US.'⁸ A European researcher explained that Beijing's interest was in two technologies in particular: those used in network-centric warfare, allowing armies to use information systems to their advantage; and smart bombs, containing the very magnets produced by Magnequench.'⁹

China's attempt to influence US ballot boxes
Any US politician who was aware of Beijing's intentions and actions knew full well that a company such as Magnequench would be a choice target for the Chinese military. One such politician was George W. Bush, who, as president of the United States, could have stopped the Valparaiso plant from going to China in 2006. But the US was busy leading a global war against terrorism; any threat not linked to Islamic fundamentalism stood little chance of capturing the attention of the White House. During the Democratic Party's Indiana presidential primary in 2008, in which Barack Obama was narrowly defeated, Hillary Clinton stated that Bush 'did nothing' to stop the demise of Magnequench. Rare metals experts quickly pointed out the hypocrisy of the claim — wasn't it the party's candidate's husband who, only a few years earlier, had allowed the negotiations that led to the buyout, despite warnings from numerous top officials at the Department of Defense?

Why was there such apathy on the part of the Democratic administration? This is where the case of Magnequench takes a troubling turn. At the time of these events, a series of technology transfers had already started to raise the hackles of the US military industry and the small circle of magnet manufacturers. One of these was Steve Constantinides. He later described the stupefaction of one of his peers upon learning that the White House had, over a period of three or four years, been divulging confidential industrial intelligence on American ballistic technologies to Beijing: 'The United States shared its industrial secrets on missile technologies with China. And Bill Clinton forced his government to do so.'[20]

What was happening behind the scenes? 'Everyone had their reasons,' replied Constantinides evasively, not wanting to say any more on the matter.[21] Leitner was more forthcoming: 'Some people argued it was because of the kickbacks and money that was given to the Clintons, the Republicans, and the Democrats ... typically from the People's Liberation Army — money from the Chinese that went directly into the White House.' But without any real evidence, such claims remain nothing more than rumours.

Subsequent documentaries co-directed by Leitner investigated the clandestine relationship between the Democratic administration and China in the 1990s. It is widely known that Beijing sought to fund the Democratic Party during the 1996 presidential elections to back Bill Clinton and his vice-presidential candidate Al Gore. Despite a US electoral law prohibiting non-American citizens from financially participating in the election process, numerous intermediaries contributed on behalf of China.

Enter the enigmatic Johnny Chung. The Sino-American citizen was so close to the Clinton couple, and his dealings so dubious, that he became known as the Democrats' key channel of funds from China. It has since been revealed that Chung gave the White House funds that he received indirectly from top officials in the Chinese army.

These facts were reported by the US press and in a documentary in which Peter Leitner was interviewed.[22] James Woolsey, the head of the Central Intelligence Agency (CIA) at the time, also agreed to be interviewed by the director, who asked him why he'd only met with Bill Clinton twice over two years, while Johnny Chung went to the White House fifty-eight times over the same period. 'Mr Woolsey could not explain it,' recalled Leitner. 'He just said that the president's schedule speaks for itself.' Chung allegedly went as far as arranging a face-to-face meeting for a high-ranking official of the People's Liberation Army — the source of the financing — with the president during a fundraising evening in Los Angeles.[23] However, the justice system quickly caught wind of the affair.[24] *The Washington Post* leaked a federal enquiry into the Chinese embassy in Washington's heavy-handed coordination of China's interference efforts. The Democratic Party was eventually ordered to return millions of dollars in donations to their many benefactors.[25]

Dozens of people, including the bagman Johnny Chung, were convicted in the massive scandal that became known as 'China Gate'. Yet the senior level of the Democratic Party was never targeted, and the attorney-general, Janet Reno, refused to appoint an independent counsel to take the investigation further. Interestingly, when President Donald Trump was faced with an in-depth enquiry by special counsel Robert Mueller into his alleged ties with Russia during the 2016 campaign, no one seemed to remember the other proven scandal involving a country infinitely more harmful to Washington's interests.[26]

Is it possible that the clandestine funding received by the Democratic camp pushed President Clinton to divulge US technology intelligence in return? It's a serious allegation. Did the White House think that the technology in question was not as strategic as experts claimed it to be, or that, with or without Magnequench's patents, the Chinese would have acquired the technology anyway? There are many grey areas that may never be brought to light. Either way, as stated by Peter Leitner, 'The Clinton administration was heavily predisposed toward approving almost anything the Chinese wanted.' Leitner also left me with an open question — one that would take months of investigation to clear up: were rare-earth magnets the essential thread in this tangle of corruption, cynicism, and a thirst for power?

South China Sea: access denied

Peter Leitner did make one thing clear: 'The Chinese targeted Magnequench in order to advance their ability to produce long-range cruise missiles.' The question is whether the technology they acquired formed part of the two ballistic missiles that were first revealed to the world during a spectacular military parade in Beijing in 2015: the Dongfeng-26 ballistic missile, capable of reaching the US base on the island of Guam, and the Dongfeng-21D anti-ship ballistic missile.

Nicknamed the 'aircraft carrier killer' and operational since 2010, the DF-21D has been central to Beijing's policy of prohibiting access to the South China Sea these past few years. Having control over this strip of ocean running from its coasts to the south of Vietnam would increase China's strategic leverage, and give it access to prodigious quantities of offshore hydrocarbon resources, as well as an eye on the comings and goings of half the world's oil.

This scenario is unacceptable to Japan, South Korea, Vietnam, and the Philippines, but especially to the US, 60 per cent of whose warships are currently positioned in the Pacific.[27] Today, barely a week goes by without a naval incident of some sort, making the territory the powder keg that could ignite a Sino-American conflict.[28]

The West's global military supremacy is unrivalled. The US accounts for 36 per cent of the world's defence expenditure, compared to China's 13 per cent.[29] And yet Beijing's capability in advanced ballistic technologies has already shifted the balance of power in the South China Sea. 'In a future conflict situation, our military is going to face an enemy that is a lot more robust, that is a lot better armed, is much more modern and technologically sophisticated, and has a great deal of more precision in their targeting capabilities than they should have,' predicted Peter Leitner. 'It means that we're going to be pushed further and further out into the Pacific.'

China is the world's second-biggest economic power, and its relationship with the US has deteriorated considerably in recent years. During his presidency, Donald Trump engaged in a fierce trade conflict with Beijing. His successor, Joe Biden, pursued this hard-line policy. In 2022, he abandoned the United States' traditional doctrine of ambiguity towards Taiwan, going so far as to state that in the event of an attack against Taiwan, the United States would defend it.[30] These statements may prompt fears of escalation, and lead to a direct confrontation in the medium term. In such a scenario, would the US not be vulnerable against an adversary that is also the source of its most critical defence components? And would China not take timely advantage of this dependence, either by playing the rare-earths card during trade negotiations, or by hampering America's military efforts?

Given their lack of interest in rare earths — and critical metals overall — over the last few decades, the US intelligence services, which are responsible for overseeing these issues, have a lot of ground to make up. In 2014, Anthony Marchese, chairman of the mining company Texas Mineral Resources, directly asked Michael Morell, the CIA's deputy director from 2010 to 2013, how involved the agency was in these issues. 'Morrell said to me then: "The good news is that the matter is in my inbox. The CIA is perfectly aware of the problem. The bad news is that it's right at the bottom of my inbox, because the White House has never indicated rare earths as a priority to the agency."'

The US government has, however, gradually begun to take action in recent years, notably from 21 July 2017, when President Trump ordered a report to assess and strengthen 'the manufacturing and defense industrial base and supply chain resiliency of the United States'.[31] Delivered in the spring of 2018, the report provided a database of the American industries that guarantee the country's military sovereignty. More importantly, it listed all the 'individual points of failure': the critical companies and factories whose extinction would paralyse the entire US defence industrial base.

Anthony Marchese again decided to dig deeper. During the same period, he set up a meeting with one of the report's authors at a coffee shop outside the White House. He learned that '[already] then, the White House didn't like the idea that the Chinese supplies the US its rare earths. So, the administration thought of introducing a "Buy American Clause" in its defence contracts.' The protectionist clause would force US defence groups to source their military components from US suppliers, which would naturally revive the domestic production and transformation of rare earths.[32]

In 2018, US secretary of commerce Wilbur Ross also announced his intention to restrict Chinese aluminium imports — far from being used for soda cans, aluminium is also a component of numerous American weapons: 'At the very same time that our military is needing more and more of the very high-quality aluminum, we're producing less and less of everything, and only have one producer of aerospace-quality aluminum.'[33]

Joe Biden religiously followed the work of the previous administration on this matter. A few months after taking office, he ordered a new audit of US security vulnerabilities with respect to critical materials. Published in June 2021, the audit found, unsurprisingly, that 'supply chains are at serious risks of disruption', and recommended that the production and processing of these resources be brought back to American soil.[34] This recommendation was taken forward the following year with federal funding to build a heavy rare-earths refining plant in Texas by the Australian company Lynas, as well as a rare-earths magnet factory, also in Texas, by the American group MP Materials.[35]

Until these initial measures to supply the US military-industrial complex with made-in-America magnets become reality, Washington can expect the same difficulties that have created thorny dilemmas for Democratic and Republican administrations — the same dilemmas, in fact, faced by the manufacturers of the revolutionary Lockheed Martin F-35 Lightning II fighter jet.

Fatal attraction: Chinese magnets and the Pentagon

The case of Lockheed Martin finds its origins in a 1973 US law that banned the procurement from foreign suppliers of specialty metals for use in military technology.[36] Lawmakers recognised that parts containing cobalt, zirconium, and titanium were increasingly important to the US Army's arsenal. America needed a domestic industrial base that could ensure the country's mineral security in times of war.

In the early 1990s, the world's leading army undertook a formidable challenge: to design and put into the sky an aircraft that could rival the French Rafale. Developed by the US defence group Lockheed Martin and co-financed by numerous US allies, the F-35 fifth-generation stealth fighter jet has already cost US$412 billion, making it one of the most expensive programs ever run by the US army.[37] America's hopes in the aircraft are as high as the tax bill for citizens: not only will the F-35 allow the US to dominate the sky, it will stimulate the country's defence industry, restore the trade balance, and create tens of thousands of jobs. Lockheed Martin produces around 150 F-35s a year to meet orders from numerous countries (the US, of course, but also Canada, Australia, the UK, Germany, Poland, Israel, and South Korea).[38]

Then, in August 2012, two of Lockheed Martin's biggest suppliers, Northrop Grumman and Honeywell, went to the White House with their concerns about the rare-earth magnets used in some of the radars, landing gears, and computer systems that they supplied for the assembly of the F-35. Northrop Grumman had discovered that its radars, installed on the 115 already-delivered stealth jets, contained magnets that were not made by a US manufacturer, but by a Chinese competitor,

ChengDu Magnetic Material Science & Technology Co. Apparently, an unscrupulous intermediary had circumvented US regulations, making the F-35 program partially unlawful and the continued purchase of these parts out of the question.

The Pentagon was alerted to the matter, which was taken up by Frank Kendall, then undersecretary of defence for acquisition, technology, and logistics. It was a dilemma: waiting for an American magnet manufacturer to provide the precious parts to replace those of the Chinese would potentially lead to delays in the F-35's deployment, and the cost of retrofitting the stealth jets with American parts to replace the offending magnets would be astronomical.

The Pentagon therefore looked into waiving the 1973 law for matters of national security. But some top officials were circumspect: how could they be sure that the rare-earth magnets China supplied to F-35 manufacturers didn't also contain spyware? Could the US$400 billion program be 'contaminated' by a few magnets costing no more than $2 a piece? By giving the Chinese control of the downstream supply of rare minerals, had the US given its competitors the opportunity to steal their military secrets and make up for lost ground?

These questions raised another broader question of national security that the US has asked itself time and again: how does it prevent the infiltration of Trojan horses in the microchips and other semi-finished goods containing rare metals sold by the Chinese around the world, including to Western armies? A 2005 report by the Pentagon even raised the possibility of electronic systems that are used extensively in US weapons being infected by malware that could disrupt combat equipment mid-operation.[39] These fears escalated when the Pentagon discovered that raw materials from China were being used in other vitally important military equipment — namely, Boeing's Rockwell B-1 Lancer long-range bomber, certain Lockheed Martin F-16 fighter jets, and all the new SM-3 Block IIA defence missiles manufactured by Raytheon.

Kendall ordered Lockheed Martin to find a solution. Until then, replacing the magnets one by one was out of the question. The technological superiority of the world's leading army and of numerous Western allies was at stake at a time when China and Russia were developing their own stealth fighter jets. Faced with severe time and budgetary constraints, and ruling out the risk of components containing backdoor technology, Kendall made his decision: the ban imposed by the law of 1973 would not apply to some of the rare-earth magnets produced by the ChengDu Magnetic Material Science & Technology Co., making the Chinese company the official supplier of the F-35.[40]

The United States cannot do without Chinese magnets. To this day, said Anthony Marchese in an interview in 2017, the Pentagon continues to grant the waiver. 'The manufacturers of the F-35 still buy rare earths in China. Period.'

That is, until the subject once again hit the headlines. In September 2022, the Pentagon outright refused to deliver new F-35s, having found that all 825 of them, manufactured by the world's leading power, contained the notorious Chinese-made magnets.[41] Since the same causes produce the same effects, it was Mr Kendall's successor, William LaPlante, who tackled the issue this time. The new deputy secretary of state again saw the inability of the US production infrastructure to supply these components to the Pentagon, and a few weeks later signed a new waiver for the F-35s already on the production line.

But there's a catch: as of 31 October 2023, which was the delivery date of the waivered F-35s, no further stealth aircraft containing Chinese magnets may be delivered.[42] Furthermore, under a law enacted on 23 December 2022, all suppliers to the US Department of Defense must, within thirty months, trace the origin of magnets containing rare earths or other rare metals at all stages of their manufacture.[43] The Pentagon now seems determined to be less reliant on its Chinese rival. And with a line-up of local magnet-manufacturing plants waiting in the wings between now and 2025 or 2026, the United States is steadily giving itself the means to implement its policy.[44]

CHAPTER EIGHT
Mining goes global

OUR NEED FOR RARE METALS IS BEING DIVERSIFIED AND expanded by green-energy generation, power transmission, and storage, the aerospace and defence industries, and now, exponentially, cryptocurrencies and generative artificial intelligence. Not a day goes by that we don't discover a new miracle property of a rare metal, or unprecedented ways of applying it. Indeed, our technological ambitions and dreams of a greener world are limited only by the bounds of our imagination. This means expanding our mining operations across the entire planet — and the Earth, we tell ourselves, will keep up. Contrary to speculation, there will always be that acre of mountain, that crease in a hill, or that clearing in a valley where we can extract those few aggregates of precious rare earths such as lithium, cobalt, graphite, and gallium. And not forgetting the more abundant metals — copper, iron, and aluminium — that are just as indispensable to new technologies.

After all, there is a precedent: between the end of the First

World War and 2007, the annual production of 14 of the minerals essential to the global economy increased 20-fold.[1] Let's shorten that time frame: between 2002 and 2015, humanity managed to rip out of the ground no less than one-third of all metals mined since the start of the twentieth century.[2] This has resulted in the rise of a host of indicators, including consumer habits, the accumulation of wealth, the number of possessions, the quantity of electronic data movements, and global warming.

Where does that leave us for the century ahead? Will this breakneck pace simply gather momentum? If global GDP grows at an annual rate of 2.5 per cent, it will have doubled by 2050.[3] By the time you read these lines, everything we build, consume, barter, and throw away will have doubled in less than a generation. There will be twice as many high-rises, highway interchanges, chain restaurants, industrial livestock-production farms, commercial aircraft, e-waste dumps, and data centres. There will be twice as many cars, connected objects, refrigerators, barbed-wire fences, and lightning conductors.

We are going to need twice as many metals, rare or not rare.

A metals shortage ahead?

Some have put numbers on our future needs. At a symposium held at Le Bourget in 2015, on the margins of the Paris climate talks, a handful of experts had already presented their forecasts.[4] They had predicted that by 2040 we will need to mine three times more rare earths, five times more tellurium, 12 times more cobalt, and 16 times more lithium than today. And the stakes are rising, as we saw with more recent figures in Chapter One: according to a 2021 report by the International Energy Agency, by 2040 humanity will have mined 42 times more lithium, 25 times more graphite, 21 times more cobalt, and seven times more rare earths than in 2020. (See Appendix 4.)[5] Olivier Vidal, a researcher at the French National Centre for Scientific Research (CNRS), even conducted a study of the metals we will need in the medium term to sustain our high-tech lifestyles.[6] His work was published in 2015, and at the time was mentioned on the BBC.[7] Over the few years that followed, he gave some 30 lectures in Europe, mostly to students.

That is as far as Vidal's study went. Yet it should be on the bedside table of every head of state the world over. Basing his research on the most widely accepted growth outlooks, he highlights the massive quantities of base metals we will need to extract from the subsoil to continue to fight against global warming. Take the case of wind turbines: by 2050, keeping up with market growth will take '3,200 million tonnes of steel, 310 million tonnes of aluminium, and 40 million tons of copper'.[8] Indeed, wind turbines guzzle more raw materials than previous technologies: 'For an equivalent installed capacity, solar and wind facilities require up to 15 times more concrete, 90 times more aluminium, and 50 times more iron, copper, and glass than fossil fuels or nuclear energy.'[9] According to the World Bank, which carried out its own study in 2017, the same applies to solar and hydrogen electricity systems, which 'are in fact significantly *more* material intensive in their composition than current traditional fossil-fuel-based energy supply systems'.[10]

The overall conclusion is aberrant. Because global metal consumption is growing at a rate of 3 to 5 per cent per year, '[t]o meet global needs by 2050, we will have to extract more metals from the subsoil than humanity has extracted since its origin'.

This bears repeating: over the next generation, we will consume more metals and minerals than in the last 70,000 years, or the 500 generations before us. Our eight billion contemporaries will absorb more mineral resources than the 117 billion humans who have walked the Earth to date.[11]

Vidal admits that the study is incomplete: assessing the actual ecological footprint of the green transition requires a far more holistic approach that includes the lifecycle of raw materials. It also requires measuring the staggering volumes of water consumed by the mining industry, the carbon dioxide emissions produced by the transportation, storage, and use of energy, the still little-known impact of recycling green technologies, and all the ways in which these activities pollute ecosystems — not to mention the extent of their impact on biodiversity.[12]

'It's mind-boggling,' admitted the researcher. And yet so few political leaders have a proper grasp of all these aspects. Vidal maintains that in recent years he tried to alert the French minister of research: 'I never made it past the first echelons of the lower administrative hierarchy.' Do the Intergovernmental Panel on Climate Change (IPCC) reports pay closer attention? 'The issue of mineral resources is still virtually non-existent', said Vidal.[13] And what about the Conferences of Parties (COP) following on from the 2015 Paris Accords? My research shows that the question of access to metals and minerals was never broached between COP 22 (2016) and COP 26 (2021). Only at COP 27, held in Sharm el-Sheikh (Egypt) in December 2022, was I invited to a panel discussion on the topic, at the initiative of several young activist organisations. And I must modestly point out that it was a 'side event', held outside the plenary discussions.[14]

Yet we are contending with a massive challenge of resource availability. On the one hand, advocates of the energy transition are adamant that we can draw infinitely on the inexhaustible sources of energy generated by the tides, the wind, and the sun to make our green technologies work.[15] On the other hand, rare metal hunters warn that we could soon run out of a considerable number of raw materials. Just as we have a list of threatened animal and plant species, soon we will have a red list of metals nearing depletion. (See Appendix 18 for the viable reserves lifespan of the primary metals needed for the energy transition.) At the current rate of production, says the French institute IFP Energies nouvelles (IFPEN), 'Almost 90 per cent of currently known copper resources may have been extracted by 2050.'[16] More pessimistic forecasts project that the resources (that is, reserves discovered, regardless of the viability of their use) of most abundant and rare metals could be depleted in less than 70 years.[17] 'Cobalt will be the next metal shortage,' predicted one electric battery-cell expert. 'No one saw this coming, and time is running out.'[18]

Will we manage to start up enough mines in the next 30 years to satisfy our appetite for metals? Our era is often referred to as the 'New Renaissance': we are at the dawn of an age of unprecedented technical invention and possibilities for exploration. Yet how can we hope to reach these new frontiers if we run out of the resources we need? What if Christopher Columbus, with no wood at his disposal, hadn't found the *Pinta* and the *Niña* caravels moored at an Andalusian port in 1492?

The energy and digital transition at stake

Diplomatic achievements, ambitious energy-transition laws, and the efforts of the most zealous environmental defenders will amount to nought without sufficient quantities of rare metals. If current data is anything to go by, the green revolution will take much longer than hoped. More importantly, it will be a green revolution led by China — one of the few countries with an adequate supply strategy. And it will not stop at producing the metals needed for the energy transition, especially for its own needs. (China alone consumes nearly 80 per cent of rare-earth oxides produced globally.)[19] It also imports metals produced elsewhere: cobalt from the Democratic Republic of the Congo, 80 per cent of which is exported unprocessed to be refined on home soil; nickel, 35 per cent of which is refined by China; and lithium, of which between 50 and 70 per cent of global production is transformed in China.[20] Beijing's first concern is quite clearly to prioritise the interests of its green-tech businesses, and to support the growth of its energy and digital transition, to the detriment of other countries.

This is how China can spurn all stereotypes of being one of the most polluted countries on the planet to instead being seen as spearheading a greener world and the fight against global warming. It would be plausible for three reasons:

- First and foremost, our collective denial of resource scarcity. Yet in 1931, the French writer Paul Valéry warned that 'the time of the finite world has begun'.[21] And in 1972 the Club of Rome presented in its pivotal report The Limits of Growth the paradox between the exponential growth of the global population and the economy, and the finiteness of resources.[22] Almost a century has passed since these first warning bells, and our behaviours have still not changed. On the contrary, the belief in an endless supply of metals prevails. Our reasoning remains akin to that of the French economist Jean-Baptiste Say who, in 1803, stated in his Treatise on Political Economy that we would not obtain natural resources for free were they not limitless.[23] And we consume even more than before. We never dreamed we would catapult ourselves into a world of scarcity the way we have. Evidently, our astounding technological transition outpaced our cognitive progress.

- A lack of investment in mining. To meet the need for electric vehicle batteries alone by 2035, 400 new mines will need to be built around the world (97 for natural graphite, 74 for lithium, and 72 for nickel.[24] All that costs money: according to Christophe Poinssot, deputy managing director and scientific director of the French Geological Survey (BRGM), '3.3 trillion US dollars will need to be invested in extraction and refining projects by 2030, but today just half of that amount has been invested'.[25] Any delay will cost us dearly in the next two decades, bearing in mind that it takes on average 16.5 years to commission a new mine.

- Added to this are the political and social constraints bound to any mining activity. An identified deposit does not an exploited deposit make. In 2021, Greenland's Inuit Ataqatigiit party, staunchly opposed to the opening of a rare-earth mine in Kvanefjeld in the south of the island by Australian company Greenland Minerals, won the legislative elections and effectively buried the mining project.[26] In 2022, Serbia bowed to public pressure by burying Europe's biggest lithium mining project, located in the Jadar region, headed by Australian mining giant Rio Tinto.[27] More recently, in the bay of Tréguennec on the far-western of edge of Brittany, France, local residents categorically refused any exploitation of a lithium deposit discovered in a Natura 2000 nature-protection area.[28]

- Compounding those limits are barriers of a geopolitical nature. Russia's invasion of Ukraine, for example, wiped out plans by Australian company European Lithium to mine two major lithium deposits in the Donbas region.[29] The lengthy process of starting up a mining operation requires political and legal stability, conditions not always met in many resource-rich developing countries. And so Afghanistan's burgeoning lithium production industry — a resource openly coveted by Beijing since the US army withdrew from Kabul in August 2021 — is likely to be a long time coming.

- And what will happen if the effects of climate change drastically reduce the water reserves needed to extract and refine minerals? In early 2022, the press reported that Australian mining group BHP would be scaling back operations at its Cerro Colorado copper mine in northern Chile following a court ruling preventing the group from extracting the blue gold from a nearby aquifer for three months.[30] And in 2023 in China, hydroelectric stations in the southern region of Yunnan also had to contend with a lack of water and to lower their power generation, forcing aluminium manufacturers to significantly reduce their output.[31]

- Lastly, the energy return on investment (EROI) — the ratio of the energy needed to produce metals to the energy generated using the same metals — is against us. Extracting one to five grams of gold requires crushing one tonne of rocks — a million times more rock than the metal itself. In the same way as a baker would probably need to mill an entire skip of bread loaves in the hope of recovering three measly cups of salt. The Italian researcher Ugo Bardi gives another example: 'Imagine that you were asked to take care of the mining waste created by the copper contained inside your new car. An average car contains about 50 kilograms of copper, mainly in the form of wiring. So, on your way home from the dealer, you would be followed by a truck that would then proceed to dump about one tonne of rocks in front of your door.'[32]

How much energy do we need to generate energy? The question may seem hare-brained for most of us, but it's top-of-mind for energy players. One century ago, extracting 100 barrels of oil required, on average, the energy supplied by one barrel of oil; today, this same barrel only produces 30 barrels of oil in some drilling areas.[33] Drilling technologies have become more efficient, but the most accessible oilfields are now depleted, and more energy is needed to reach new and harder-to-access reserves. For non-conventional crude (shale oil and oil sands), one barrel will produce four barrels at the most.[34] We are teetering on the absurd! Will our production model still be sound when it takes one barrel to fill another?

The same applies to rare metals, which require increasing amounts of energy to be unearthed and refined. Experts rightly state that there are more rare-mineral deposits to be discovered than currently proven, so there is no need to worry about coming up short.[35] But producing these metals takes 7 to 8 per cent of global energy.[36] What if this ratio were to jump to 15 to 20 per cent, or more? It's a deeply legitimate question. 'Compared to 1900, the mining industry today uses sixteen times more power to produce a tonne of copper', Mark Cutifani, then chief executive of British mining company Anglo American, told us in a conversation in Paris in 2021. 'We also move sixteen times more tonnes of ore and waste to produce the same amount of copper.'[37] As for gold, between 1998 and 2017 the pure-metal content of certain mines in the state of Nevada operated by Canadian company Barrick Gold fell from 12.04 to 3.02 grams per tonne of ore.[38] Is a future where these figures continue to climb sustainable?

The same can be demonstrated with just about every mining

resource. Some cases are so critical that a deposit containing as many minerals as it did in the 1980s is now considered a 'diamond in the rough' in the mining world.[39] As the researcher Ugo Bardi concludes, 'The limits to mineral extraction are not limits of quantity; they are limits of *energy*.'[40]

We are starting to see the limits of our production system; they will be reached the day we need more energy to produce the energy we need. This is in tandem with the many more constraints listed above, which, as I understand it, are not necessarily geological, but of a much more diverse nature — political, economic, geopolitical, social, ecological, even psychological! This is why Fatih Birol, executive director of the International Energy Agency, tentatively asserted in 2021 that 'the data shows a looming mismatch between the world's strengthened climate ambitions and the availability of critical minerals that are essential to realising those ambitions'.[41] This is also why Simon Michaux, geologist at the Geological Survey of Finland (Geologian Tutkimuskeskus, GTK), stated outright in 2023 that the energy transition 'will not be possible for the entire global human population' and that 'there is simply just not enough time, nor resources to do this by the current target set by the world's most influential nations'.[42]

Yet our instinct for conquest forever compels us to keep pushing the envelope to expand humankind's domination in every nook and cranny of the globe (going as far as outer space, as we will discover).

I had to find out more. I took a train to London to peruse ancient maps in the hope that they would provide useful information to assuage our appetite for green growth.

Counter-attack from the West and the multiplication of mines

Awaiting me at the Geological Society of London, on the banks of the river Thames, was 'the map that changed the world'.[43] For nearly two centuries, the precious parchment has slumbered in the depths of the GSL archives. To get there, I had to walk through the stately entrance of Burlington House — an imposing building complete with a neo-Renaissance façade overlooking the Piccadilly high street. On the first floor, at the end of a threadbare carpeted staircase, there is a patio with walls lined with old books that serves as a reading room. In the light of two ageing chandeliers, archivist Caroline Lam delicately pieced together fifteen square pages, each one measuring 60 centimetres long and wide. Together they form a treasure measuring 3 metres by 4 metres — one of the first detailed mining maps in history.

The Great Map is the work of William Smith. Over ten years in the early nineteenth century, the geologist surveyed Great Britain on foot and on horseback with the aim of describing the mineral lie of the land. The copy that is kept at the Geological Society of London is one of the first to have been printed in 1815 — the year in which the map was presented to the public. I needed a magnifying glass to make out the names on the map. Easier to identify were the colour-coded minerals that are plotted in all their diversity: chalk and sand quarries alongside limestone and marble deposits. Drawn in black are the coal seams that would make Great Britain vastly wealthy throughout the nineteenth century.

At the time William Smith published the Great Map, Great Britain was in the throes of its industrial revolution. In the mills, steam generated thermal energy and powered the spinning jennies that would significantly increase productivity. The same steam power led to the introduction of the locomotive on an increasingly dense railway network, which in turn contributed to the phenomenally swift expansion of trade and progress. But to actuate the locomotive pistons and set its wheels in motion, the steam needed to reach a temperature of close to 350 degrees Celsius. Boilers were therefore fitted with coal furnaces.

This fossil fuel soon became a highly prized resource. 'People needed to know where the deposits were,' said Caroline Lam as she painstakingly put away the fragments of the Great Map. With the help of Smith's map, miners rushed to the coal seams to supply the fuel for Great Britain's new energy requirements. In this respect, the Great Map well and truly transformed the world by stimulating the first industrial revolution and giving Great Britain a head start on the rest of Europe. In the Victorian age, the nation used its dominant position in coalmining to establish its industrial, technological, and military superiority, and to become the world's superpower.

Two centuries later, we want to apply the British example to the energy and digital revolution: it will take a new Western sovereignty over our metal supplies to circumvent China's grip. And so, Europe and the United States are laying the groundwork for a counter-attack ... If anything, the geopolitical context of late has reinstated the question of strategic autonomy, starting with the Covid-19 pandemic, when we discovered — to our astonishment — that our dwindling supplies of masks came from China.[44] Then, in March 2021, the six-day blockage of the Suez Canal by the container ship Ever Given laid bare the vulnerability of the global supply chain when one of the links

jams.[45] Last but not least, in June 2022, Moscow turned off the
tap on its gas exports to Europe — a heavily reliant importer.[46]
All these crises explain why, in 2023, President Macron spoke of
'European sovereignty' during his speech in The Hague on the
future of the continent.[47]

On the metals front, a first — and spectacular — response
came from the US in 2022, when Congress voted in the Inflation
Reduction Act.[48] The purpose of this formidable raft of legal and
tax measures is to incentivise the reshoring of critical mineral
and green-technology production to American soil. That same
year, President Joe Biden invoked the Defense Production Act
— an act dating back to the Cold War that gives him access
to federal funds to support the extraction, processing, and
recycling of critical metals.[49] The European Union wasn't far
behind: in her State of the Union address on 14 September 2022,
European Commission President Ursula von der Leyen declared
— finally! — that '[l]ithium and rare earths are already replacing
gas and oil at the heart of our economy'.[50] Then, in March 2023,
the European Commission presented the Critical Raw Materials
Act — a plan to revive the extraction, refining, and recycling
of critical raw materials on European soil, and to diversify
the European Union's sources of supply. France was not to be
outdone either: a report outlining courses of action to secure
the supply of mineral raw materials to industry was submitted
to the government in 2022 by Philippe Varin, president of the
Nancy World Materials Forum.[51]

Also in 2022, France launched the French Observatory of
Mineral Resources (Ofremi) to monitor the critical metal needs
of French industry. States are also finetuning and updating their
lists of critical resources. In 2023, the European Commission

published its updated list, which now includes 34 resources (compared with 30 in 2020). US authorities have raised their number of minerals at risk of supply disruption to 50 (up from 35 in 2018). Such audits are just a prelude to unbridled prospecting for potential mining partners: governments, carmakers, and other entrepreneurs in the scramble for metals. This is being undertaken not only at a national scale, as in Smith's time, but on a planetary scale: deposits of rare earths have been discovered in at least 29 countries on five continents. (See Appendix 19 for the world map of rare-earth mines and deposits).[52] North Korea is believed to have some of the most abundant rare-earth deposits in the world, and mining companies have already started exploring the hundreds of rare metals deposits around the globe.[53]

The trend is anything but sound, with speculative bubbles bursting as mining companies admit that some deposits yield far less than initially announced. Like a spin of the roulette wheel, fortunes are made by the minute while small-time players lose their nest eggs overnight. Either way, the frenzy is creating geopolitical upheaval that goes against the comradely ideals displayed at the conclusion of the Paris Accords.

Countries are therefore striking up new alliances for rare metals exploration. The United States actively practises mining diplomacy in the Democratic Republic of the Congo and Zambia;[54] German Chancellor Olaf Scholz visited Canada in 2022 to sign mining partnerships;[55] France is prospecting for critical metals in Mongolia and Chile;[56] Tokyo is teaming up with Delhi, Canberra, and Washington to develop new rare-earth refining technologies;[57] and in 2022, the US launched the multilateral Partnership for Mineral Security with a dozen countries (including Australia, France, Japan, and Finland), all committed to sourcing minerals according to shared environmental, social, and governance standards.[58]

The rise of the 'electro-state'

This diplomatic activism could also give rise to new mining powerhouses from the emerging world: Chile, Peru, and Bolivia, thanks to their abundant lithium and copper reserves; India, with its rich titanium, steel, and iron reserves; Guinea and southern Africa, whose subsoils are packed with bauxite, chromium, manganese, and platinum; Indonesia, where nickel and tin are abundant; and New Caledonia, thanks to its generous nickel deposits, etc.

Better still, in response to new Western appetites, these states want to do more than simply sell their raw minerals, as we saw in previous chapters. They want to sell their added value as well. They have understood that only by moving up the value chain — as China did before them — will they emerge the winners of the energy transition. And so, in 2022 Mexico's Andrés Manuel López Obrador announced the nationalisation of the country's lithium;[59] Brazil's Luiz Inácio Lula da Silva and Chile's Gabriel Boric want to increase mining taxes and royalties;[60] Ghana's Nana Akufo-Addo refuses to export unprocessed lithium;[61] and Indonesia's Joko Widodo plans to build gigafactories on home soil.[62]

Of course, an integrated chain cannot be built by decree, and there are tremendous human, financial, infrastructural, political, and other challenges to consider. But what we need to understand here is that our thirst for lithium, cobalt, copper, and rare earths will not be easily quenched by countries that have understood that the era of the petro-state is giving way to that of the 'electro-state'.[63] Chile, the Democratic Republic of the Congo, Indonesia, and Peru are set on becoming such states, and are structuring part of their influence accordingly.

The mining customer will no longer (always) be king. 'Given the increased competition between consumer countries, it will be less a case of importers deciding to buy metals, than it will be of producers deciding to sell metals,' a French expert confirms. 'We are going to have to come to terms with this new concept of "competitive consumption."'[64] In this respect, the success of French mineral diplomacy will depend, I believe, on the promise of a fair share of the windfall, and a transfer of value and skills from the North to the South, as the British weekly *The Economist* suggests.[65] It's a call being made loud and clear in Asia and Africa. On a visit to the Democratic Republic of the Congo in 2022, US Secretary of State Anthony Blinken made this very promise, and the message was positively received by Zambia, in particular.[66]

The energy and digital transition is sending humanity on a quest for rare metals, and is upending traditional balances of power. Rather than abating the geopolitics of energy, it is compounding them.[67] It is a new world that China wants to fashion to its liking by embarking on its own quest for rare metals in Argentina, Australia, Greenland, Myanmar, and New Caledonia. But the most prized location of all is Africa — home to 30 per cent of the world's mineral reserves.[68] From its bauxite mining projects in Guinea and Ghana, lithium projects in Zimbabwe and Namibia, and buying up future graphite production in Mozambican and Tanzanian mines,

China's efforts are proving far more assertive than those of the West. The result is that Africa is now the primary beneficiary of Chinese companies' exploration budgets. As much as 7 per cent of Africa's mining production is believed to be controlled by Chinese interests.[69] The case of the Democratic Republic of the Congo is particularly telling: at last count, Chinese-backed or -owned companies hold stakes in 15 of the 19 cobalt mines in the country, the world's leading producer of the resource.[70] And overall, nearly 70 per cent of the DRC's mining activity is controlled by Chinese companies.[71]

With its mining-expansion strategy, the Middle Kingdom is working towards a bold objective: to abandon the mining monopolies built on domestic mineral resources in favour of a new dominant position, by controlling the production of a bounty of rare metals across the planet. It's as if Saudi Arabia, which holds the largest proven reserves of oil worldwide, took it upon itself to control the oil reserves of the now 13 members of OPEC.

Most Western governments still have little idea of the scale of the economic war being waged by China in the subsoil of developing countries. But there is one downstream segment of the metals value chain where Paris, Berlin, Washington, and Ottawa have been redoubling their efforts in recent years: the ever so strategic industry of electric-vehicle batteries.

Battle of the batteries
The manufacture of electric batteries — accounting for at least 30 per cent of the total price of the vehicle — is a critically important venture for Europe and the United States. Their

combined automotive sectors employ over 20 million people (13 million in Europe and 7 million in the US as of September 2022). But accounting for 77 per cent of the world's lithium-ion battery production today is China — as I discovered in 2019, while reporting from the southern Chinese city of Shenzhen, home to the electric-vehicle production lines of the formidable Chinese brand BYD (Build Your Dreams). The message delivered by its communications director, Richard Li, was clear: the automotive group's command of the rare metals supply chain is allowing it to hit its production targets and become a world leader in the sector.[72] This reality also led Simon Moores, director of the lithium battery price information agency Benchmark Mineral Intelligence, to comment in 2020 that, 'China is building one battery gigafactory a week; the US one every four months'.[73]

But a counterattack is taking shape, with Europe supporting its industrial champions and attracting those from the American and Asian continents to keep green jobs on its territory. In Dunkirk, northern France, Taiwanese group ProLogium has started building a battery gigafactory, operational from 2026; Swedish company Northvolt is commissioning a gigafactory in northern Germany in the same year; and in Europe, the Chinese giant CATL (Contemporary Amperex Technology) is building its gigafactory in Hungary in 2023 to meet the needs of European carmakers Mercedes-Benz, BMW, Stellantis, and Volkswagen. (Production is scheduled to start in 2025.) All told, some 50 gigafactories are expected to open on the continent by 2030.

The US's Inflation Reduction Act, which incentivises manufacturers to set up their operations there, has proved a fantastic accelerator. In June 2023, for example, Ford received a staggering US$9.2 billion in loans from the federal government to set up three plants in the states of Kentucky and Tennessee. Japan's Honda and South Korea's LG received US$4 billion to build a gigafactory in Ohio. At this rate, by 2030 China's share of global battery production will decline slightly to stabilise at 60–65 per cent while Europe's share will increase to 20–25 per cent and North America's to 15–20 per cent.[74]

Also instrumental to Western resilience is its grasp of the chemistry of the materials needed for future generations of electric-vehicle batteries. The aim is to 'leapfrog' a generation of technologies by using new raw materials with physical and chemical properties that are potentially superior to lithium, cobalt, and nickel, thereby regaining its advantage over China. Most of today's batteries are of the 'NMC' (nickel-manganese-cobalt) variety, but a competing technology known as 'LFP' (lithium-iron-phosphate), or simply the lithium-ion battery, is growing, equipping up to one-third of the world's electric cars to date. Cheaper and longer-lasting, these batteries have the advantage of doing without critical resources such as cobalt and nickel. Other innovations, such as lithium-sulphur and lithium-air batteries (with a theoretical energy density of two to four times that of a lithium-ion battery), are being researched. Another expected breakthrough is the sodium-ion battery, which does not need lithium and would cost half as much as the batteries currently on the market. These new batteries could be manufactured in France by the French company Tiamat by 2025.[75]

And what about batteries made from food waste fibres? In 2023, researchers at Italy's institute of technology (IIT — Istituto Italiano Di Tecnologia) developed an entirely edible battery made from seaweed, almonds, capers, and beeswax.[76] In Switzerland, scientists have designed a battery made from paper,[77] while the University of Maryland (US) has invented a battery made from crab and lobster shells![78] While we can't expect such batteries to equip millions of electric vehicles any time soon, they are confirmation that the 'energy revolution' is chiefly a raw materials revolution, and that the energy transition will not be based on the materials and technologies we know today.

So what will be the strategic materials of the future? Will cobalt and nickel retain their prestige on the American and European criticality lists over the next two decades? There are as many opinions as there are experts, some arguing that the breakthrough of sodium batteries — a mineral wildly abundant in the Earth's crust — will sound the death knell for lithium, condemning it to the annals of history. Other experts, however, point to an unrelenting trend in history: from one technological revolution to the next, scientific innovations become more complex, and use an ever-wider range of elements from Mendeleev's periodic table of elements. In the days when armies fought their battles with cannons made of bronze, the strategic resources were most certainly copper and tin — the essential metals of this alloy. While the military has moved on from bronze in the 21st century, copper is essential for electricity networks, and tin for microprocessors. So while the use of metals may evolve, their strategic importance remains unchanged.

Can we not then make the same observation about the rare metals needed for batteries? Will the strategic nature of lithium and nickel be augmented (rather than replaced) by other resources with properties yet unknown? If this is the case, our criticality indexes could, in the not-too-distant future, be updated for very abundant resources such as iron, sodium, and even certain bio-sourced waste products. And perhaps our governments will give greater attention to recycling such materials.

The geopolitics of the circular economy

In the West, there is growing awareness of the need to replenish the strategic stocks squandered after the collapse of the USSR. This is what the European Critical Raw Materials Act sets out to do, by coordinating EU member states in creating stocks of certain critical resources.[79] And in 2022 in the US, the issue of strategic stocks was removed from the sole purview of the US Department of Defense, where it had long been confined; the US departments of energy, defense, and state signed a memorandum of understanding laying the groundwork for a joint stockpile of minerals to be made available to the defence and energy transition industries in the event of a future trade crisis.[80]

What remains is the colossal challenge of reducing our consumption of resources. It bears repeating that the energy transition is making us realise that 'cleaner' driving and heating our homes in a 'greener' way means digging deeper; that emitting less carbon dioxide into the atmosphere means producing more metals and minerals. Like motorists waiting at a level crossing, behind the first train that passes before us — the low-carbon revolution — lurks another: the circular revolution: a consumption model where resources are reused, rather than thrown away, to limit our impact on ecosystems. We need to pursue not one, but *two* revolutions if we are to achieve a truly sustainable world. And to implement an ambitious circular economy we would need to reduce metal extraction by one-third by 2050.[81] The reality, however, is that we're doing just the opposite: the switch to all-electric only corroborates the recent OECD forecast that humanity will consume 2.5 times more resources in 2060 than it did in 2011.[82]

Thus, realising the circular economy is a Copernican task; it involves using less material at every stage in the lifecycle of a resource, and keeping it in the economy as long as possible. It is a revolution that would require recycling the water needed for mining at the metal-extraction phase; eco-designing the technology to facilitate their recycling downstream; building cooperation between industries that collect waste and those that reuse it; and growing the rental and sharing economy for consumer goods, which would logically reduce their production. We also need to extend the lifespan of objects, especially batteries, and systematically recycle everything that can be recycled. (See Appendix 20 for the seven pillars of the circular economy.)

The circular economy is not just for the sake of the environment; it is an unavoidable step for governments and companies on the long road to mineral independence. A circular world would consolidate the resilience of Western states by doubling the sectors of primary resources (from mining) with sectors of secondary materials (from recycling plants), thereby diversifying supply. Herein lies the truly geopolitical dimension of recycling: for the countries and companies that will benefit from the resource; and for the countries that will refuse to export it in the future, for example by imposing quotas and embargos on exports of used batteries, as is already the case today for virgin raw materials.

Under the European Critical Minerals Act there is nevertheless a target of 15 per cent of critical metals consumed each year to be recycled by 2030. Better still, the new batteries regulation adopted in July 2023 by the European Council requires recycled products to be reused. And by 2031, new electric-car batteries will have to contain 16 per cent recycled cobalt and 6 per cent recycled lithium and nickel.[83] This is an exciting target, but no mean feat: most of the 'recyclable' lithium in circulation is in the batteries of brand-new electric vehicles, which won't be on the scrap heap for another 10 to 15 years! We can build all the recycling plants we want right now, but they will stand idle for lack of available waste. And while we wait for these materials to come rolling in, we'll need to rely on mining for many years to come.

And then by the time the first generation of electric cars reach their end-of-life stage in one or two decades' time, our need for lithium will have exploded. But the quantity of resources available for recycling in 2035 or 2045 will never exceed that which was integrated into batteries 10 or 20 years ago — that is, five to 20 times less than our new needs. This means that for a long time to come, compared with virgin raw materials, recycled critical metals will make up a very slim portion of our consumption of resources. In short, we won't stop putting holes in the Earth's crust any time soon.

This obvious fact stems from a simple observation: with our economies growing at 3 per cent a year, our progress on the circular economy front, while real, falls short of our increasing consumption of materials. Every year, more virgin metals enter the economy than second-hand metals, which has led the consultancy firm Deloitte, in its Circularity Gap Report 2023, to find that from 9.1 per cent in 2018, the circularity of the global economy dropped to 8.6 per cent in 2020, and stood at 7.2 per cent in 2023, 'driven by rising material extraction and use'.[84] Recycling may well be a long-term strategy, but it is by no means an immediate solution for alleviating the pressure on resources as a result of our modern lifestyles, and weaning us off China in the short term.

While we wait for the highest levels of governments and environmental groups to truly grasp the real complexity of the 'green' world, Beijing is pursuing its strategy of encircling the West. The latest episode took place on 3 July 2023: after several Western countries imposed restrictions on the export of electronic-chip technology to China, China announced that from the following month, the export of gallium and germanium — two critical metals for the manufacture of microprocessors and fibre-optic technologies, and of which it is the world's leading producer — would be subject to government approval.[85] It's a response that exposes yet again our dangerous addictions.

We understand that our addiction to Chinese metals will be long-term. That is, unless France, and other Western countries and their allies decide to open rare metal mines on home soil.

CHAPTER NINE
The last of the backwaters

SHOULD WESTERN COUNTRIES BE REINSTATED AS MINING powers? The idea is inconceivable for most, sickening even. In France, environmentalists are against the idea, but President Emmanuel Macron is not.

The idea has been tossed around within the French government for years. It was in one of its glass offices overlooking the Seine that Arnaud Montebourg, the minister of industry under François Hollande, voiced his intention to reopen France's mines: 'The renewal of mining in France is underway ... We want to make sure our country is supplied with the raw materials it needs to ensure its independence, cost and quantity control, and sovereignty.'[1]

France: a slumbering mining giant
Montebourg's words echoed the reindustrialisation policy promised by Hollande during his 2012 presidential campaign.

The thinking is sound, for France is in fact a mining giant lying dormant (see Appendix 21).[2] Until the early 1980s, France had a strong mining industry for minerals as diverse as tungsten, manganese, zinc, and antimony, beginning in the first industrial revolution in the nineteenth century. It also had a prosperous iron industry, following the founding of the European Coal and Steel Community (ECSC) in 1952. France's booming mining industry generated thousands of direct and indirect jobs, from the Maurienne valley in the Alps to the rugged landscape of Lorraine in the north-east, and from the Black Mountain folds in the southern centre to the Mouthoumet massif further west. France was on its way to becoming one of the world's biggest antimony, tungsten, and germanium producers.

To turn words into action, Montebourg had proposed the creation of a national company of mines, the CMF. It would have a budget of up to €400 million to invest in mining companies, explore partnerships in Africa, and deliver mining permits in mainland France, and thus 'reengage France in the global battle for access to natural resources'.[3]

But following the financial mishaps of two French mining houses, Eramet and Areva (which became Orano in 2018), Montebourg's successor, Emmanuel Macron, finance minister from 2014 to 2016, shelved the project: 'Eramet and Areva were under strain after commodity prices crashed. It would be unwise to create a third fully state-owned enterprise. We prefer to focus on current restructuring.'[4]

The setback didn't stop Macron — once a member of parliament, and now holding the country's highest office — from launching in 2015 a 'responsible mining' project to reduce the environmental impact of all future mining projects,[5] nor from issuing some 20 metals and mining research permits in metropolitan France and French Guiana.[6] In 2021, the government also enacted a reform of the mining code with a view to making mining activities on French soil compatible with environmental protection. In 2022, the reform of the code was completed with a number of decrees by including mining damage in the scope of environmental and health damage, and by extending the penalties that could be imposed by the Police des Mines.[7] Finally, in 2023, the French National Centre for Scientific Research (CNRS) and the Geological Survey (BRGM) launched a €71.4 million seven-year exploratory research programme called 'The Subsurface as a Common Good', with the aim of improving overall knowledge of France's geological resources.[8]

Europe is also on the ball. Presented in 2023, the European Union's Critical Minerals Act states that by 2030, 10 per cent of Europe's critical resource needs must be met by mines operating in the continent's own subsoil. Dozens of mining projects have in fact been announced in recent years. Battery-grade graphite might be mined in Sweden from as early as 2024,[9] as well as rare earths in the longer term, as recently announced by the Swedish mining group LKAB following the discovery of what could be the largest deposit on the European continent.[10] In Portugal, the authorities have given the go-ahead for the development of the country's first lithium mine, in the north-east of the country;[11] and in Germany, the Australian company Vulcan Energy is planning to develop lithium in the geothermal waters of the Upper Rhine Valley from 2027. (See Appendix 22 for lithium mining projects in Europe.)[12]

The question of restarting mining across the European

continent has fundamentally changed the nature of the debate. Until now, it was easy to criticise Beijing for manipulating raw-material prices and ignoring international trade rules. But as Europe looks to revive its own mining industry, it must face up to its own responsibilities.

In France, the government's position on the controversial topic has polarised public opinion in recent years, and sparked citizen movements and resident associations across the country who have taken to the streets, chanting, 'Not here, not anywhere.'[13] One of the environmental groups is the highly vocal Les Amis de la Terre (Friends of the Earth), which has labelled the government's promise of sustainable ore exploration as 'unrealistic', and has denounced the 'mistruths' of the mining revival.[14] In 2022, the announcement in the local press in Brittany that the subsoil of Tréguennec, a town in France's far-western Finistère department, contained a deposit of 66,000 tonnes of lithium also aroused strong opposition from local residents.[15] Also in 2022, the Imerys group unveiled plans to open a lithium mine in the Allier region of France by 2027 to manufacture batteries.[16] This prompted the association France Nature Environnement to express fears about the impact of such an operation on biodiversity, water resources, and air quality.[17]

And who can blame them? Numerous mining disasters in France have hardened the population against the industry.[18] Western citizen movements have gone from NIMBY — Not In My Back Yard — to BANANA — Build Absolutely Nothing Anywhere Near Anything.[19] The situation isn't any easier in the rest of Europe. Public protests in Serbia, for example, led the government to revoke, in 2022, the licences granted to miner Rio Tinto for a lithium project in the western region of Jadar.[20]

And in Sweden, Canadian miner Tasman Metal's plans to mine rare earths at the Norra Kärr site in the south of the country has for years been the subject of both institutional and public opposition.[21]

But these groups, often forerunners to such protests, are fundamentally inconsistent. They condemn the effects of the very world they wish for. They do not admit that the energy and digital transition also means trading oilfields for rare metals deposits, and that the role of mining in the fight against global warming is a reality we have to come to terms with.[22] French government press releases state as much: reopening French mines 'is part of the national ecological strategy towards sustainable development'.[23] The debate is also opening our eyes to what the Chinese have known all along: the Western development model is mired in contradictions. It is a hard choice between dreams of a greener world and the reality of a more technological world.

All the makings of an anti-mining rebellion are in place, and could delay a mining revival in France. For Europe, it's hard to project a medium-term outlook. Philippe Varin, author of a report on critical metals commissioned by the Elysée Palace, published in part in 2022,[24] believes that the continent will be able to secure at best 20 per cent to 30 per cent of its needs by 2030.[25] As for the entire rare metals industry, as early as 2010 the US Government Accountability Office had forecast that it would take a good fifteen years at least to rebuild the industry.[26] And while Western countries wait for stakeholders to reach an agreement, their mining culture is struggling to take root. Training is insufficient, and young people are no longer drawn to careers in geology. As the last of the talents disappear, there is a real risk that the sector's revival may be decades in the making.

I support bringing back mining in the West. Not so much for the value, the additional tax revenues, and the thousands of jobs it would create; nor for the strategic security of having our own supply chain at a time when producer countries are tightening the noose around our necks. Rather, my argument is on behalf of the environment.

Reopening mines in the West is the best possible decision we can make for the health of the natural environment. Relocating our dirty industries has helped keep Western consumers in the dark about the true environmental cost of our lifestyles, while giving other nation-states free rein to extract and process minerals in even worse conditions than would have applied had they still been mined in the West, without the slightest regard for the environment.

The effects of returning mining operations to the West would be positive. We would instantly realise — to our horror — the true cost of our self-declared modern, connected, and green world. We can well imagine how having quarries 'in our backyard' would put an end to our indifference and denial, and drive our efforts to contain the resulting pollution. Because we would not want to live like the Chinese, we would pile pressure onto our governments to ban even the smallest release of cyanide, and to boycott companies operating without the full array of environmental accreditations. We would protest en masse against the disgraceful practice of the planned obsolescence of products, which results in more rare metals having to be mined, and we would demand that billions be spent on research into making rare metals fully recyclable. Perhaps we would also use our buying power more responsibly, and spend more on eco-friendlier mobile phones, for instance. In short, we would be so determined to contain pollution that we would make astounding environmental progress and wind back our rampant consumption.

In this scenario, China's mining activities would face real competition from more ethical Western mines. If China thus lost its edge, its soil, rivers, and air would be beneficiaries. Consumers would be better informed and more demanding, and the industry's only chance at winning back market share would be to improve its own practices. China's environment would emerge the winner from this virtuous competition imposed by the West.[27]

Nothing will change so long as we do not experience, in our own backyards, the full cost of attaining our standard of happiness. Mining the earth responsibly on our own turf will always be more valuable than mining irresponsibly elsewhere. It would be a deeply ecological, selfless, and brave decision that adheres to the ethics of responsibility extolled by countless environmental groups — and rightly so.

An example was the protest in 2009 against European countries exporting their nuclear waste to Russia.[28] Some activists went as far as tying themselves to railway tracks to stop trains coming out of waste-storage facilities, protesting that we alone are responsible for processing our waste, and that palming it off to others is downright immoral.[29] This kind of virulent opposition to the way the end of the fuel cycle is managed should focus just as much on how our dirty mining is outsourced. Protesters should be forming human chains around the ports of Le Havre, Algeciras, or Rotterdam to stop shipments of metals from China from entering the European Customs Union, and should chain themselves to the gates of their government buildings until a law is passed that allows rare earths mining in their countries.

Paris and the conquest of the seas

In France, activists should have also asked the government to keep a closer eye on the actions of two Pacific island kings, Filipo Katoa and Eufenio Takala. These enigmatic royals control a wealth of rare earths so great that the future of France's energy and digital transition may depend on their good graces.

On 12 July 2016, wearing the traditional *maros* over their suits, the monarchs from Alo and Sigave — located in the Wallis and Futuna French island collectivity in Far Oceania between Tahiti and New Caledonia — mounted the steps of the Élysée Palace to attend a reception in their honour. While in Paris they were shown around the National Assembly, the Senate, and the Ministry of Overseas Territories.[30] Together with their delegation, King Filipo Katoa and King Eufenio Takala had come to discuss matters such as the opening of their kingdoms, the economic dynamism of their region, and better access to healthcare. It's an important relationship, as shown in 2016 when President Hollande invited the kings to his presidential stand alongside the prime minister of New Zealand, John Key, and the US secretary of state, John Kerry, for the Bastille Day parade. Paris is keeping the two royals in its inner circle while it works out how to maintain its presence on their isles. Will it take increasing the €15 million it already spends every year on this special relationship? And will France manage to keep any irredentist hankerings at bay, the consequences of which would be disastrous for the country?

The 16,000 kilometres separating the capital, Mata Utu, from Paris makes the Polynesian kingdoms the furthest French territory from the mainland since becoming French protectorates in 1887. Yet despite this, nothing in Wallis and Futuna happens according to French standards. Decentralisation laws, international treaties, land regulations, and the traffic code all take second place to local custom, which is guarded by the 'last kings of France' and their mystical powers.[31] Even trade unionists may not strike without first seeking the approval (or ensuring the neutrality) of the 'great chiefdom' of Kings Katoa and Takala.[32]

The kings do not have absolute power, but share it with tribal institutions and with the custodian of antediluvian protocol, the chief of ceremony. The French government maintains the status quo with a monthly salary of €5,500 for the two kings.[33] Indeed, 'Wallis and Futuna are republican kingdoms,' said Pierre Simunek, a former sub-prefect in Mata Utu.[34]

Another infringement of French rule in Wallis and Futuna is the partial application of the 1905 French law on the separation of the church and the state. The basic public service of primary education is entrusted to the church, which by tradition is a powerful institution in Wallis and Futuna.[35] What's more, territorial assemblies are usually opened with a prayer by Bishop Susitino Sionepoe.[36] 'The sessions start with a blessing from the Holy Spirit, everyone makes the sign of the cross ... and then the shouting begins,' recalled Simunek.[37]

The former sub-prefect shared another telling detail. Above the assembly president's seat is a portrait of the French president. Above his portrait are the flags of the Wallis and Futuna kingdoms, which are in turn below the French flag. And topping the entire arrangement is a large crucifix. To Simunek, this implied that Wallis and Futuna's prefect and sub-prefect — the highest representatives of a country known for its aversion to religious symbols — are in this context 'protectors of the throne and altar'.

To what lengths is France prepared to go to maintain its sovereignty over a territory barely bigger than the city of Paris? If France is kowtowing, it is because the isles allow France to stretch its footprint into the Pacific. It gives Paris a say over the world's biggest ocean and largest trade area, and enables it to weigh in on negotiations with regional organisations (such as the Pacific Islands Forum), and to build partnerships with de facto neighbours — and with Australia in particular. As stated in France's white paper on defence: 'New Caledonia and the communities living in French Polynesia and Wallis and Futuna make France a political and maritime power in the Pacific.'[38] In 2023, the military programming law was adopted to strengthen the resources of the French navy to consolidate the protection of overseas territories.[39]

The main reason, however, is that Wallis and Futuna give France exclusive access to what in the region is known as 'La Grande Marmite' — the great cauldron. The 20-kilometre-diameter crater left over from the Kulo Lasi volcano was discovered following three exploration campaigns between 2010 and 2012.[40] For years, the French Geological Survey (BRGM), the French Institute for Integrated Marine Science Research (IFREMER), and the Eramet mining group had been trying to assess the mineral potential of the kingdoms' exclusive economic zone (EEZ), believed to contain a priceless treasure trove of manganese, nickel, copper, cobalt, zinc, lead, and possibly rare earths.

Its discovery has created huge tensions in Wallis and Futuna. Fearing that Paris will stake its claim, the chiefdoms are claiming ancestral rights on the territory both above and below water, and Futunans, together with Wallisian elected officials, have demanded the immediate suspension of exploration campaigns. 'One of the ministers of the Royal Counsel even threatened secession over rare earths,' said Simunek.[41] And the visit to Wallis and Futuna in 2018 by four scientists from the Research Institute for Development (IRD) specialising in the exploration and study of seabed minerals showed French authorities just how touchy the subject was, with the chiefdoms and local populations again staunchly opposing any exploration and mining of their subsoils.[42]

Wallis and Futuna are not the only isles with mineral-rich lagoons: the exclusive economic zones of Tahiti and Clipperton Island, in the north-east Pacific, also have a wealth of submerged rare metals.[43] Other states, such as Japan, Papua New Guinea, and New Zealand's Cook Islands have made similar discoveries in the Pacific and Atlantic oceans. (See Appendix 23 for a map of deep-sea rare-metal resources.)[44] It would seem that entire swathes of marine areas — 71 per cent of the Earth's surface — are more than watery wastelands where shoals of fish spawn. The 'blue economy' has the potential to generate exponential wealth.

As a new goldrush forms on the horizon, the rare-earths battle (like that of the energy and digital transition) is taking to the seas. By June 2023, around 30 exploration licences had been issued, by Japan and Norway in their exclusive economic zones, and on the high seas by the International Seabed Authority (ISA).[45] Spearheading the offensive are three private companies:

Canada's The Metals Company, Belgium's Global Sea Mineral Resources (GSR), and Norway's Loke Marine Minerals.[46] Always on the ball, China has designed submersibles capable of exploring the ocean floor at record depths. 'Beijing has positioned itself using financing the West doesn't have,' explained a marine geosciences expert. 'The exploration of the oceans has only just begun.'[47]

France is also leading the pack. It has successfully executed its maritime-extension policy in the last few years by applying the international law of the sea, defined in the Geneva convention of 1958, to encroach on the international maritime areas adjacent to land surfaces under its control, including Guiana, Martinique, Guadeloupe, New Caledonia, and the Kerguelen Islands. France's maritime domain now spans over 11.7 million square kilometres — twenty times mainland France's surface area — making it the world's biggest territory, followed by the US (11.3 million square kilometres).

Even those nostalgic for its lost colonial empire cannot deny just how big the French republic is today. And it could get even bigger: the Commission on the Limits of the Continental Shelf (CLCS) — the UN body which sets the outer limits of coastal countries — is considering extending the underwater areas owned by France to 350 nautical miles (650 kilometres), provided France can demonstrate that these are natural extensions of its land surface.[48]

France is not the only country to stake its claim on the maritime *Monopoly* board. Canada, Denmark, Australia, Russia, Japan, Côte d'Ivoire, and Somalia are just some of the dozens of countries requesting the extension of their exclusive economic areas. The Russian Federation has its eyes on part of the Eurasian basin of the Arctic Ocean; Brazil, on an area of the Atlantic Ocean close to the equator; Kenya, on a large area of the Indian Ocean; the Republic of Mauritius, on the Rodrigues Islands region; Nigeria, on an area of the Gulf of Guinea; the Palau Islands, on an area on the northern shelf of the archipelago; Portugal, on an area off the Azores and Madeira, as well as on an area off Galicia, jointly with Spain. Sri Lanka and India each want areas of the southern Bay of Bengal; France and South Africa want to share the continental shelf in the region of the Prince Edward Islands and the Crozet archipelago (both in the Indian Ocean).[49] Some countries have even resorted to subterfuge: China has gone as far as building artificial islands in the South China Sea so that it can claim exclusive use of the surrounding marine territory.[50]

In short: for thousands of years, 71 per cent of the planet's surface did not belong to anyone; for 60 years, countries owned 40 per cent of the surface of the oceans, and a further 10 per cent is now the subject of continental shelf-extension requests. We can surmise, therefore, that coastal states have jurisdiction over 57 per cent of the seabed.[51] The allure of rare metals has led to the biggest 'land' grab in history, and in record time.

But there is also something heartening in what we're seeing. For thousands of years, humans the world over have impaled, stabbed, and disembowelled each other over land that makes up one-third of the planet. Now we are deciding on another one-third of the planet — half of the oceans — in no time at all and without having to kill anyone. Only battalions of lawyers armed to the teeth with international law are required. It is significant progress, which proves that humanity is capable of improving over time.

The downside, however, is that the exponential growth of our need for rare metals could increasingly commoditise the world's backwaters, which have long been spared from humanity's greed. Case in point, the head of The Metals Company (Canada), Gerard Barron, has very openly stated that there is enough cobalt and nickel in seafloor polymetallic nodules to power the batteries of around 4.8 billion electric cars.[52]

The ecological challenges are as unprecedented as they are poorly understood. What impact will undersea mining activities have on ecosystems? 'A few experiments were carried out on nodule fields in the Clarion-Clipperton zone in the 1970s and 1980s. Forty years later, their ecosystems have still not fully recovered from these disturbances', warns a chemist and specialist in seabed ecosystems at the French Institute for Integrated Marine Science Research (Ifremer).[53]

Sensing the danger, at COP 27 Emmanuel Macron called for a moratorium on mining in international waters beyond national jurisdiction.[54] His stance was in line with that of the European Parliament, which in 2018 also came out in favour of a moratorium on deep-sea mining.[55] It also aligns with that of 57 parliamentarians from 32 countries who, in the summer of 2022, called for the suspension of such activities.[56] Will this be enough to safeguard the last vestige of biodiversity? Nothing could be less certain: in 2021, Canadian miner The Metals Company, sponsored by the island nation of Nauru (Pacific Ocean), applied to the ISA for authorisation to mine rare metals in the Clarion-Clipperton zone. With no mining code in place to govern this operation, The Metals Company could launch its assault on the deep seabed within two years, prompting negotiations that were held at ISA headquarters in Kingston, Jamaica, in July 2023. At the time of writing, a General Assembly of 168 States was debating, in relative secrecy, the philosophy of the energy transition. Which of the countries opposed to deep-sea mining (such as Chile, France, Germany, and Ireland) and those that support this extension of the mine (including China, Russia, South Korea, and Norway) will emerge triumphant? Whatever the outcome — and pending the international community's official stance on this thorny issue — states are already divvying up the sea like parcels of land.[57]

The day President Obama fired the starting gun on a new space race

Outer space isn't out of bounds either. This is despite the 1967 Outer Space Treaty, which clearly states that the space beyond the ozone layer is the common property of humanity. But like the oceans, humans are already preparing for minerals from outside the Earth's atmosphere.

The United States was first off the mark. In 2015, President Obama signed off the revolutionary Commercial Space Launch Competitiveness Act, which grants any US citizen the right to 'possess, transport, use, and sell' any space resource. The wording is subtle; it does not openly challenge the international law establishing the principle of non-ownership of celestial bodies, but claims the right of ownership of the riches they bear.[58]

It's a nuance that changes the way we survey the sky. Capitalism — by which objects are ascribed value — has apparently given rise to a new breed of gold-panner who sees asteroids as bags of cash in orbit. Just as a kilogram of apricots goes for a few dollars at the market, an acre of land in Montana or Queensland is worth thousands of dollars, and Modigliani's *Reclining Nude* is auctioned off for US$170 million, an asteroid crossing the Earth's orbit could sell for thousands of billions of dollars.

Since 2015, Americans have started to put price tags on these celestial juggernauts. One of them, the 2011 UW-158, which narrowly missed the Earth's surface in its namesake year, was estimated to be worth €4,000 billion.[59] It was crammed with 90 million tonnes of rare metals — the 'new oil' of the energy and digital transition — including more platinum than humans have ever extracted from the Earth's crust. Then, in 2021, the media picked up on another potato-shaped asteroid, dubbed '16 Psyche', containing iron, nickel, and gold with a total estimated value of $10,000 million trillion — more than the global economy.[60] By opening the door to ownership of space minerals, President Obama has given the self-proclaimed 'space gold-panners' of Silicon Valley the legal assurance they need to act on their mining ambitions. These companies include Californian start-ups such as AstroForge and TransAstra, which have announced space exploration missions with a view to mining the Moon and asteroids.[61]

For the time being, commercial space exploitation is utopian at best. The cost of space launches is prohibitive, and there is no ecosystem in place for companies to profit from off-Earth activities. So why didn't the international community simply laugh off the 2015 Space Launch Act? Because space ownership is not a question of how, but of when. So, while member states of the European Space Agency address the taboo concept of space mineral ownership under Obama's Space Launch Act, international agencies will have to start laying down the legal and diplomatic framework for divvying up the sky.[62] In 2022, the UN Committee on the Peaceful Uses of Outer Space

(COPUOS) set up a working group for a period of five years to assess if the 1967 International Space Treaty should be revised and how.[63]

And for good reason: to ensure access to ownership as part of 'New Space' — the fledgling private space economy that has come with the emergence of US space entrepreneurs (Elon Musk of SpaceX; Jeff Bezos of Blue Origin; Greg Wyler of OneWeb;, and E-Space, an American start-up that wants to launch 10,000 nanosatellites into space to bring the Internet of Things to life). Enter other players like Luxembourg, which is already securing its position.[64] In 2016, then finance minister Etienne Schneider announced the Asteroid Mining Plan — the first European space initiative to promote a favourable legal framework for asteroid mining. In 2019, he signed an agreement with the US to share intelligence on space. There is even the provision of €200 million in funding to incentivise space-mining companies to set up in the Grand Duchy. The tax haven is focusing on new growth opportunities by fashioning itself as the global hub of the New Space Economy so that it can attract entrepreneurs and jobs … and generate handsome tax revenues.[65]

We need to think very hard about the moral of this story of land, oceans, and asteroids. The more equal distribution of resources that we celebrate has, in fact, led to the biggest drive for mineral ownership the world has ever seen. The ambition to reduce humans' impact on the ecosystem by way of the energy and digital transition has actually further trampled biodiversity underfoot. And now, our new-found craving for minerals in space is wiping out the last of the sacred wildernesses. Are we looking to the skies to invoke the gods — or subjugate them?

Epilogue

IN THE MID-NINETEENTH CENTURY, WHALE OIL WAS AS essential as fossil fuels are today. The first industrial revolution in Europe brought the need for better lighting. Lamps using vegetable oils, mineral oils, and animal fat were until then the best way to conquer darkness. Then humans became hooked on whale oil, whose handsome flame would light up households and public streets both effectively and inexpensively. Soon fleets of fortune-seeking whalers were crisscrossing the oceans in search of millions of gallons of the precious fat.

This was the gold rush that created the whaling industry. The sector produced 40 million litres of oil a year, and sparked wars in the Sea of Japan and in the North Pacific over control of prime whaling areas. But so many whales were slaughtered that hunting became increasingly difficult, making oil harder to come by, and hiking the cost of lighting.

Having managed this resource so poorly, would humans learn to do without their precious lighting? Not in the least. In 1853, the Polish pharmacist Ignacy Łukasiewicz developed a lamp that used a lighter, more functional oil: kerosene. Petroleum would become the next ideal fuel, until electricity become commonplace in the twentieth century.

For many historians and economists, there is a lesson to be learned from our reckless quest for whale oil. Rather than reassess our need for lighting, as our short-sightedness should have taught us, we found a way to illuminate our lives even more by way of petroleum, and from the resilience and prosperity it offered. It's a lesson we need to remember in the twenty-first century as we witness the emergence of many new and abundant energies. Scientists are proposing the implementation of laser and magnetic-confinement fusion, hydrogen-powered and magnetic-levitation vehicles, and even solar power stations placed in the Earth's orbit.[1]

Green technologies will also improve: work is underway to replace the silicon in solar panels with much cleaner and more efficient photovoltaic cells made of a mineral compound called perovskite, and to reduce by two-thirds the carbon dioxide emissions generated by the manufacture of electric batteries.[2] We will also undoubtedly make significant headway in electricity storage, and develop new materials with revolutionary properties. Myriad innovations may render warnings from environmentalists null and void by proving, yet again, that every time an energy source approaches depletion, we have managed to replace it with another.[3] The innovation that saves us from the darkness and confirms the resilience of our species unremittingly wards off what Irish playwright George Bernard Shaw refers to as the 'tragedy' of desire.

But we can draw another lesson from whale oil. The crisis resulting from its depletion 150 years ago forced us to rethink the way we consume. Yet little came of this introspection, for history repeats itself as new resources run out with every change in the energy model.

And this cycle is unlikely to end anytime soon. Today's new energy technologies will also draw on new raw materials, both natural and synthetic. Polymers, nanomaterials, co-products from industrial processes, bio-based products, and fish waste will become part of our daily lives. We will also turn to hydrogen and thorium, in turn generating their share of environmental waste. Third-generation biofuels sourced from the far reaches of arid deserts and the depths of the oceans will be refined, using highly complex chemical processes. Cooking oil, animal fat, and citrus zest will be collected, using energy-intensive logistical networks. Millions of hectares of forest will be felled and transformed in sawmills of titanic proportions.

The resources of the future will bring new, protean challenges. The question we need to ask ourselves now is: what is the logic behind this next technological leap we all embrace? Can we not see the absurdity of leaping into an environmental sea change that could poison us with heavy metals before we have even seen it through? Can we seriously advocate Confucian harmony through material wellbeing if it means the very opposite: new health problems and environmental chaos?

What is the point of 'progress' if it does not help humanity progress?

Albert Einstein left us with a powerful statement: 'We cannot solve our problems with the same thinking we used when we created them.' Only with a revolution of consciousness can an industrial, technical, and social revolution be meaningful.

This book has sketched out sparse evidence of such

leaps of consciousness in the rare metals industry: German manufacturers opting for more expensive tungsten to maintain the diversity of their supply; attempts by Chinese authorities to end the rare-earths black market in Jiangxi province so as to protect the resource; and in Tokyo, Professor Okabe's attempts to recycle metals using salt from the high plains of Bolivia.

For their part, consumers can do more through their own behaviour. The awareness is there, and every one of us already recognises the need to limit our consumption of electronic goods built for obsolescence, to 'eco-design' goods for easy recycling and less waste, to opt for short supply loops, and to focus on saving resources.[4] While moderate consumption does not necessarily lead to 'degrowth', the best energy is that which we use wisely.[5]

I end on this note with French engineer Christian Thomas, who leaves us with a comment of optimism and common sense: 'We don't have a rare material problem; we have a grey matter problem.'[6]

Will we know how to put our grey matter towards finding the antidote to rare metals?

Acknowledgements

IT'S EASY TO THINK THAT WRITING A BOOK IS A LONELY endeavour. But it is, in fact, the result of collective work built on contributions, discussions, criticism, and encouragement from numerous sources. It speaks to the interest (often), enthusiasm (sometimes), and kindness (always) of the friends, colleagues, and specialists who honoured me with their involvement. In acknowledging their contribution, which spans several years in some cases, I am also sketching out the landscape of my professional and private life. In particular, I would like to thank:

Hubert Védrine, for writing the preface of this book after reading, annotating, and raising questions, and sometimes contradictions. Our discussions were of inestimable value.

Jean-Paul Tognet, for his honest and sincere accounts, as well as his willingness to read and reread these pages, giving his feedback over the phone from the Ile de Ré.

Christian Thomas, who, from his Paris office adorned in African masks, impressed on me the importance of rigorous analysis.

Paul de Loisy, who, during hours of rich and fascinating discussion on the terrace of a Paris café, helped enhance the book's contents.

Axel Robine, who between flights meticulously and generously read my manuscripts.

Camille Lecomte, who shared her analyses with me, even though we didn't always agree on them!

Didier Julienne, Jack Lifton, and Christopher Ecclestone, who were kind enough to share their considerable expertise, in France, Canada, London, and the United States.

Philippe Degobert, who reviewed the updates on electric motors.

Pierre Simunek, who patiently initiated me in the mysterious ins and outs of Wallisian political life.

Cookie Allez, who helped me mature as a writer.

The Pijac team, and its lifelong chairperson, for their support.

Hélène Crié and Yvan Poisbeau, who provided me with the necessary documents on Rhône-Poulenc.

Randy Henry and the LightHawk team, who gave me access to a light aircraft to fly over the deserts of California and Nevada.

The France-Japan Press Association and Scam, for their financial support of this editorial adventure.

Gérard Tavernier, who set up a number of invaluable meetings.

Félicie Gaudillat, who helped put together a long and comprehensive bibliography.

Stéphanie Berland-Basnier, for her legal advice.

Muriel Steinmeyer, for her loyalty.

Céline Gandner, who initiated the introductions that would set many things in motion.

My sister, Camille Pitron, who supported me in a way only she knows ... and many, many more!

Bibliography

Books

Aggeri, Franck, *L'Innovation, mais pour quoi faire ? Essai sur un mythe économique, social et managerial*, Seuil, 2023

Bardi, Ugo, *Extracted: how the quest for mineral wealth is plundering the planet*, Chelsea Green, 2014

Barré, Bertrand and Bailly, Anne, *Atlas des énergies mondiales: quels choix pour demain?*, Autrement, 3rd edition, 2015

Beffa, Jean-Louis, *Les Clés de la puissance*, Seuil, 2015

Bergère, Marie-Claire, *Chine: le nouveau capitalisme d'État*, Fayard, 2013

Bihouix, Philippe, *The Age of Low Tech: towards a technologically sustainable civilization*, Bristol University Press, 2020

Bihouix, Philippe, *Le bonheur était pour demain. Les rêveries d'un ingénieur solitaire*, Points, 2022 (1st edition, Seuil, 2019)

Bihouix, Philippe and de Guillebon, Benoît, *Quel futur pour les métaux? Raréfaction des métaux: un nouveau défi pour la société*, EDP Sciences, 2010

Bilimoff, Michèle, *Histoire des plantes qui ont changé le monde*, Albin Michel, 2011

Bougon, François, *Inside the Mind of Xi Jinping*, C Hurst & Co, 2018

Carton, Malo and Jazaerli, Samy, *Et la Chine s'est éveillée. La montée en gamme de l'industrie chinoise*, Presses des Mines, 2015

Castaignède, Laurent, and Bihouix, Philippe, *Airvore ou le mythe des transports propres. Chronique d'une pollution annoncée*, Écosociété, 2022

Chaline, Éric, *Fifty Minerals that Changed the Course of History*, Firefly Books, 2012

Chalmin, Philippe (dir.), *Des ressources et des hommes*, Nouvelles Éditions François Bourin, 2016

Chancel, Claude and Le Grix, Libin Liu, *Le Grand Livre de la Chine*, Eyrolles, 2013

Cohen, Élie, *Le Colbertisme high-tech. Économie des télécoms et du grand projet*, Hachette Livre, coll. 'Pluriel', 1992

de la Croix, Séverine, Pitron, Guillaume, and Lavoine, Jérôme, *Prométhium. Librement adapté du livre La Guerre des métaux rares*, Massot éditions / Les Liens qui Libèrent, 2021

Debeir, Jean-Claude, Deléage, Jean-Paul, and Hémery, Daniel, *Une histoire de l'énergie*, Flammarion, 2013

Delcourt, Laurent, *Transition 'verte' et métaux 'critiques'*, Éditions Syllepse, coll. 'Alternatives Sud', 2023

Deneault, Alain and Sacher, William, *Imperial Canada Inc.: Legal Haven of Choice for the World's Mining Industries*, Talonbooks, 2012

Dufour, Jean-François, *Made by China. Les secrets d'une conquête industrielle*, Dunod, 2012

Flipo, Fabrice, *L'impératif de la sobriété numérique. L'enjeu des modes de vie*, Éditions Matériologiques, 2020

Flipo, Fabrice, Dobré, Michelle and Michot, Marion, *La Face cachée du numérique. L'impact environnemental des nouvelles technologies*, L'Échappée, 2013

de Gaigneron de Marolles, Alain, *L'Ultimatum. Fin d'un monde ou fin du monde?*, Plon, 1984

Giraud, Pierre-Noël and Ollivier, Timothée, *Économie des matières premières*, La Découverte, coll. 'Repères', 2015

Guillebaud, Jean-Claude, *Le Commencement d'un monde. Vers une modernité métisse*, Seuil, 2008

Harari, Yuval Noah, *Sapiens: a brief history of humankind*, HarperCollins, 2015

Izraelewicz, Erik, *L'Arrogance chinoise*, Grasset, 2011

Kaku, Michio, *Physics of the Future: how science will shape human destiny and our daily lives by the year 2100*, Anchor, 2012

Kara, Siddharth, *Cobalt Red: how the blood of the Congo powers our lives*, St. Martin's Press, 2023

Kempf, Hervé, *Fin de l'Occident, naissance du monde*, Seuil, 2013

Kempf, Hervé, *Le mur de l'Ouest n'est pas tombé*, Pierre-Guillaume de Roux, 2015

Laws, Bill, *Fifty Plants that Changed the Course of History*, Firefly Books, 2010

Le Moigne, Rémy, *L'Économie circulaire: comment la mettre en oeuvre dans l'entreprise grâce à la reverse supply chain?*, Dunod, 2014

Lenglet, François, *La Guerre des empires*, Fayard, 2010

Lenglet, François, *La Fin de la mondialisation*, Fayard, coll. 'Pluriel', 2014

Meadows, Donella H., Meadows, Dennis L., Randers, Jorgen, and Behrens III, William W., *The Limits to Growth: a report for the Club of Rome's project on the predicament of mankind*, Universe Books, 1972

Mousseau, Normand, *Le Défi des ressources minières*, Multi-Mondes Éditions, 2012

Mouton, Servane (co-ord.), *Humanité et Numérique. Les liaisons dangereuses*, Éditions Apogée, 'Les Panseurs sociaux' coll., 2023

Pitron, Guillaume, *The Dark Cloud: how the digital world is costing the earth*. Translated by Bianca Jacobsohn. Scribe, 2023

Rabhi, Pierre, *The Power of Restraint*, Actes Sud, 2018

Rifkin, Jeremy, *The Third Industrial Revolution: how lateral power is transforming energy, the economy, and the world*, Palgrave Macmillan, 2011

Rifkin, Jeremy, *The Zero Marginal Cost Society: the internet of things, the collaborative commons, and the eclipse of capitalism*, Palgrave Macmillan, 2015

Roger, Alain and Guéry, François (dir.), *Maîtres et protecteurs de la nature*, Champ Vallon, 1991

Sanderson, Henry, *Volt Rush: the winners and losers in the race to go green*, Oneworld Publications, 2022

Schmidt, Eric and Cohen, Jared, *The New Digital Age: reshaping the future of people, nations, and business*, Knopf, Random House Inc., 2013

Surendra, M. Gupta, *Reverse Supply Chains: issues and analysis*, CRC Press, 2013

Tingyang, Zhao, *The Tianxia System: an introduction to the philosophy of world institution*, Nanjing, Jiangsu Jiaoyu Chubanshe, 2005

Valentin, Michaël, *The Tesla Way: the disruptive strategies and models of Teslism*, Kogan Page, 2019

Valéry, Paul, *Regards sur le monde actuel*, Librairie Stock, Delamain et Boutelleau, 1931

Winchester, Simon, *The Map that Changed the World: William Smith and the birth of modern geology*, HarperCollins, 2001

Essential reading: reports

'Study on the Critical Raw Materials for the EU 2023 (final report)', European Commission, 2023

'Mineral Commodity Summaries 2023', United States Geological Survey (USGS), January 2023

'The strategic issues concerning rare earth elements and strategic and critical raw materials', report by Patrick Hetzel and Delphine Bataille for the French Parliamentary Office for Science and Technology Assessment (OPECST) Office, no. 617, t. II, 2015–2016, 19 May 2016

'Mining and Metals. A power base for all nations. Locus of mining 1850-2030', RMG Consulting, Stockholm (Sweden), January 2021

'The geopolitics of the European Green Deal', Policy Contribution no. 04/21, Institut Bruegel / European Council on Foreign Relations (ECFR), 2 February 2021

'Second Continental Report on the implementation of Agenda 2063', African Union Development Agency — NEPAD, February 2022

'The growing footprint of digitalisation', Foresight Brief no. 027, United Nations Environment Programme, November 2021

'Democratic Republic of the Congo: Time to recharge. Corporate action and inaction to tackle abuses in the cobalt supply chain', Amnesty International, 15 November 2017

'The Role of Critical Minerals in Clean Energy Transitions', International Energy Agency (IEA), May 2021

'Metals for Clean Energy. Pathways to solving Europe's raw materials challenge', Eurometaux / KU Leuven, April 2022

'Strategic raw materials for defence. Mapping European industry needs', The Hague Centre for Strategic Studies, January 2023

'Report on international ocean governance: an agenda for the future of our oceans in the context of the 2030 SDGs', Report A8-0399/2017, European Parliament, 18 December 2017

'The Global E-waste Monitor — 2017, United Nations University (UNU), International Telecommunication Union (ITU) & International Solid Waste Association (ISWA)', report by C.P. Baldé, V. Forti, V. Gray, R. Kuehr, and P. Stegmann, Bonn/Geneva/Vienna.

Essential reading: journals

'Critical metals, can ethics and sovereignty be reconciled?', Geosciences no. 26, French Geological Survey (BRGM), June 2022

'Mineral resources: Rare earths', Feature article, French Geological Survey (BRGM), 12 July 2022

Essential reading: articles

'The transition to clean energy will mint new commodity superpowers', *The Economist*, 26 March 2022

'Commodity markets: strategies needed!', *Paris Tech Review*, 14 May 2012

'How the world depends on small cobalt miners', *The Economist*, 5 July 2022

'Métaux rares : ces entreprises lancées dans la course aux abysses', *Le Monde*, 6 January 202.

'China dominates the rare earths market. This US mine is trying to change that', *Politico*, 14 December 2022

'Batteries: "L'Europe est dépendante de la Chine non seulement pour les matières premières, mais aussi pour les produits semi-finis"', *Le Monde*, 15 January 2023

'Why China could dominate the next big advance in batteries', *The New York Times*, 12 April 2023

'Seabed mining: Growing calls for a "precautionary pause"', *Le Monde*, 10 July 2023

Essential viewing

Lepault, Sophie, and Franklin, Romain, *The New World of Xi Jinping*, ARTE, 2021

Mönch, Max, and Lahl, Alexander, *Ocean's Monopoly*, Werwiewas, 2015

Noirfalisse, Quentin, and Zajtman, Arnaud, *Cobalt Rush: the future of going green*, Les Films du Tambour de Soie, Dancing Dog Production, Esprit libre Production, Belgium, 2022

Perez, Jean-Louis, and Pitron, Guillaume, *The Dark Side of Green Energies*, Grand Angle Productions, 2020.

Pitron, Guillaume, *Terres rares, le trésor caché du Japon* [Rare Earths, Japan's Hidden Treasure — French only], Mano a Mano, 2012

Pitron, Guillaume, and Turquier, Serge, *Rare Earths: the dirty war*, 2012

Secrets of the Super Elements, presented by Mark Miodownik, BBC, 2017

Tison, Coline and Lichtenstein, Laurent, *Datacenter: the hidden face of the web*, Camicas Productions, 2012

Useful online resources
MINERALS, METALS, AND MINING

National institutes, specialised laboratories and institutions

British Geological Survey (BGS): bgs.ac.uk

Critical Materials Innovation Hub of the Ames National Laboratory (US DOE Energy Innovation Hub): www.ameslab.gov/cmi

European Commission, Raw materials, metals, minerals and forest-based industries page: single-market-economy.ec.europa.eu/sectors/raw-materials_en

French Geological Survey (BRGM): brgm.fr/energy

GeoRessources Laboratory of Université de Lorraine: https://georessources.univ-lorraine.fr/en

German Federal Institute for Geosciences and Natural Resources: gr.bund.de/EN

OECD Due Diligence Guidance for Responsible Supply Chains of Minerals from Conflict-Affected and High-Risk Areas: oecd.org/corporate/mne/mining.htm

Raw Materials Information System (RMIS), the materials research and innovation platform of the European Commission: rmis.jrc.ec.europa.eu

United States Geological Survey (USGS): usgs.gov

Professional bodies

Eurometaux (European industry association of non-ferrous metals producers and recyclers): eurometaux.eu

International Tin Association: internationaltin.org

Shanghai Metal Market (SMM): metal.com

United States Magnetic Materials Association: usmagneticmaterials.com

Specialist and expert information websites

Blog of Didier Julienne, non-ferrous metals sourcing specialist: didierjulienne.
eu/english/

Investor Intel, market news with critical minerals and rare earths coverage:
investorintel.com/category/critical-minerals-rare-earths

MINING.COM, online magazine on mining and the metals industry:
mining.com

World Materials Forum, monthly newsletters: worldmaterialsforum.com/
monthly-letters.html

Energy transition, renewable energies, and electro-mobility

Fraunhofer Research Institution for Battery Cell Production (FFB): ffb.
fraunhofer.de/en.html

French Agency for Ecological Transition (ADEME): ademe.fr/en

French Alternative Energies and Atomic Energy Commission (CEA),
Cadarache Centre — New Energy Technologies: cadarache.cea.fr/cad/
Recherche/technologie-energie

International Energy Agency: iea.org

International Renewable Energy Agency (IRENA): irena.org

IFP Energies nouvelles (IFPEN), institute of research and training in the
fields of energy, transport, and the environment: ifpenergiesnouvelles.
com

New Energy and Industrial Technology Development Organization (Japan):
nedo.go.jp/english

The 'solid-state chemistry and energy' Chair, Collège de France, addressing
materials and technologies used for batteries: www.college-de-france.

fr/fr/chaire/jean-marie-tarascon-chimie-du-solide-et-energie-chaire-statutaire

Digital technology

EcoInfo, a services group under the French National Centre for Scientific Research (CNRS) focused on the social and environmental challenges arising from information and communication technologies: ecoinfo.cnrs.fr

Electronic Communications, Postal and Print media distribution Regulatory Authority of France (Arcep): en.arcep.fr/

GreenIT, a collective of 'digital sobriety' and eco-design experts: greenit.fr

Protecting the environment and human rights

Official bodies

French Institute for Radiation Protection and Nuclear Safety (IRSN): en.irsn.fr

French National Institute for Industrial Environment and Risks (Ineris): ineris.fr/en

The Basel Convention for controlling transboundary movements of hazardous wastes and their disposal (UN Environment Programme): basel.int

Non-profits

Amnesty International, international NGO defending human rights: amnesty.org

French Commission for independent research and information on radioactivity (Criirad): criirad.org

Friends of the Earth International, non-profit dedicated to protecting the environment, the climate, and to defending human rights: foei.org

Global Reporting Initiative, independent body promoting sustainability reporting standards: globalreporting.org

Greenpeace, international NGO dedicated to protecting the environment: greenpeace.org (international)

World Wide Fund for Nature (WWF), NGO dedicated to protecting the environment, and to sustainable development: wwf.org

International and strategic policy, geopolitics, and defence

Carnegie Europe, a think tank on international policy. Refer to the Economics, Trade, and Energy research area: carnegieeurope.eu

Center for strategic and international studies (CSIS). Its research programme focuses on energy security and climate change: csis.org/programs/energy-security-and-climate-change-program

Foundation for Strategic Research (FRS), a French centre of expertise on international security and defence, 'energy and primary goods' research team: https://frstrategie.org/en/fields/energy-and-primary-goods

French Institute of International Relations (IFRI): ifri.org/energy and the IFRI Center for Energy & Climate: ifri.org/en/recherche/thematiques-transversales/energie-climat

Institut for Strategic Research of the French Ministry of Armed Forces (Irsem): irsem.fr/en

Royal Institute of International Affairs (RIIA), more commonly known as Chatham House: chathamhouse.org

Stockholm International Peace Research Institute (Sipri): sipri.org and its Military Expenditure Database: sipri.org/databases/milex

Seas and oceans

Bloom, a non-profit founded by the eco-activist Claire Nouvian, dedicated to protecting the ocean: bloomassociation.org/en

Commission on the Limits of the Continental Shelf (CLCS): un.org/depts/los/clcs_new/clcs_home.htm

International Seabed Authority (ISA): isa.org.jm

National Institute for Ocean Science (Ifremer): en.ifremer.fr

The Metals Company, specialised in offshore metals mining: metals.co

The Ocean Technology Campus, University of Rostock (Germany): oceantechnologycampus.com/en

World Wide Fund for Nature (WWF), NGO dedicated to protecting the environment, and to sustainable development. See the page dedicated to protecting the deep seabed against mining: wwf.panda.org/discover/our_focus/oceans_practice/no_deep_seabed_mining

Space

European Space Agency (ESA): esa.int

National Aeronautics and Space Administration (NASA) : nasa.gov

ONERA, the French aerospace lab: onera.fr/en

The United Nations Office for Outer Space: unoosa.org

Institutional websites

France

France Stratégie, institution aimed at stimulating public debate and clarifying collective decisions on social, economic, and environmental matters: strategie.gouv.fr/english-articles

OPECST, the Parliamentary Office for the Evaluation of Scientific and Technological Choices: www.assemblee-nationale.fr/dyn/16/organes/delegations-comites-offices/opecst

European Union

Joint Research Center (JRC), the scientific and technical research laboratory
    of the European Union: joint-research-centre.ec.europa.eu

International

African Development Bank Group: afdb.org

UN Environment Programme: unep.org

World Bank: worldbank.org

World Economic Forum, Centre for Advanced Manufacturing and Supply
    Chains: centres.weforum.org/centre-for-advanced-manufacturing-and-
    supply-chains

World Economic Forum, Centre for Energy and Materials: centres.weforum.
    org/centre-for-energy-and-materials

World Trade Organization: wto.org

Economic analyses and data

Bruegel, European think tank specialised in the economy: bruegel.org/
    topics/green-economy

Global Trade Alert, an independent organisation monitoring economic
    policies and their impact on global trade: globaltradealert.org

London Metal Exchange (LME), the leading global market for industrial
    metals: lme.com/Metals

Databases

Database of the World Bank: data.worldbank.org

Database of the French Agency for Ecological Transition (ADEME)
    providing official public data on emission factors and carbon
    accounting: base-empreinte.ademe.fr

Organisation for Economic Co-operation and Development (OECD) geographic database of raw material trade: compareyourcountry.org/trade-in-raw-materials

OECD data and statistics on the economics of material resources: doi.org/10.1787/data-00695-en

European Geological Data Infrastructure (EGDI): www.europe-geology.eu

Eurostat, the statistics bureau of the European Union: ec.europa.eu/eurostat

Mineral Resources Online Spatial Data, the mining resources database of the United States Geological Survey (USGS): mrdata.usgs.gov

Appendices

Appendix 1

Periodic table of elements

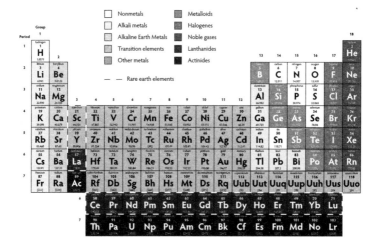

SOURCE: IUPAC, 2022.

Appendix 2

Trends in world primary metal production

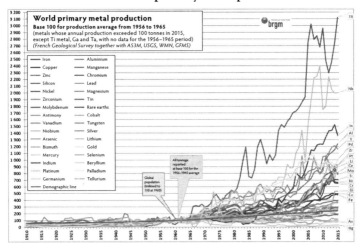

SOURCE: The French geological survey (BRGM), 2015.

Appendix 3

History of rare-earth elements production

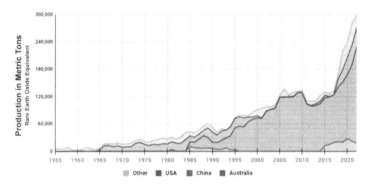

SOURCE: Hobart M.King, *REE. Rare Earth Elements and their uses*, Geology.com, 2022.

Appendix 4

Projected growth in rare metal and mineral needs (2040 relative to 2020)

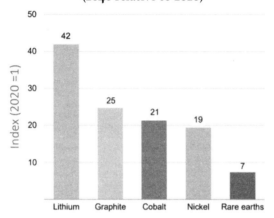

These forecasts correspond to a transition scenario that limits global warming to two degrees. Demand for lithium will increase 42-fold between 2020 and 2040, and 7-fold for rare earths.

SOURCE: The Role of Critical Minerals in Clean Energy Transitions, International Energy Agency (IEA), 2021.

Appendix 5

World map of the main producers of critical minerals and metals

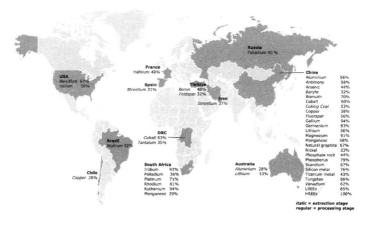

SOURCE: Study on the Critical Raw Materials for the EU 2023. Final report, European Commission, 2023.

Appendix 6

Overview of the metals used in an electric vehicle

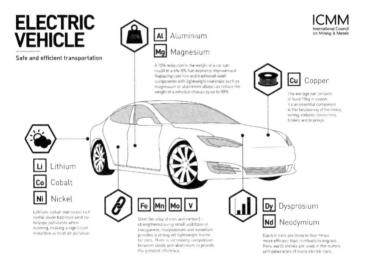

SOURCE: International Council on Mining & Metals (ICMM), 2021.

Appendix 7

Material intensity of vehicles and sources of electricity

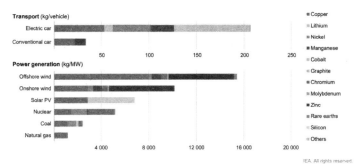

Notes: kg = kilogramme, MW = megawatt. Steel and aluminium not included. See Chapter 1 and Annex for details on the assumptions and methodologies.

SOURCE: The Role of Critical Minerals in Clean Energy Transitions, International Energy Agency, 2021.

Appendix 8

Rare-metal composition of a smartphone

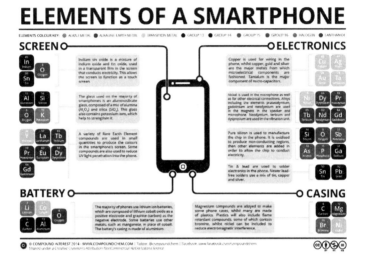

SOURCE: 'The chemical elements of a smartphone', Compound Interest, 19 February 2014.

Appendix 9

The lifecycle of metals

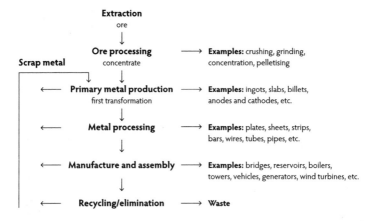

SOURCE: Gouvernement du Québec, Ministère de l'Énergie et des Ressources naturelles, 'Guide de redaction d'une étude d'opportunité économique et de marché pour la transformation au Québec', The Ministry of Energy and Natural Resources of the Government of Quebec, October 2015, p. 1.

Appendix 10

Summary table of metals recycling rates

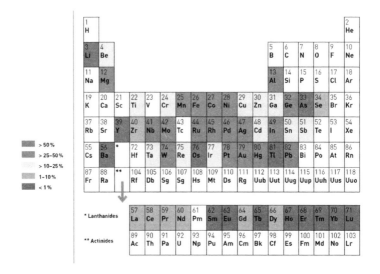

SOURCE: Recycling Rates of Metals: A Status Report, United Nations Environment Programme, 2011.

Appendix 11

The lifecycle of metals

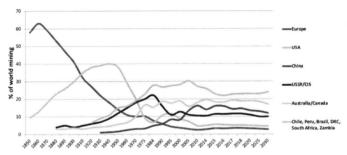

Locus of world mining 1850-2030
metals and industrial minerals
(per cent of value of total world production excluding coal)

Source: RMG Consulting 2021

DRC: Democratic Republic of the Congo

SOURCE: Mining and Metals. A power base for all nations, RMG Consulting, Stockholm, Sweden, March 2021.

Appendix 12

China's projected share of global demand for raw materials

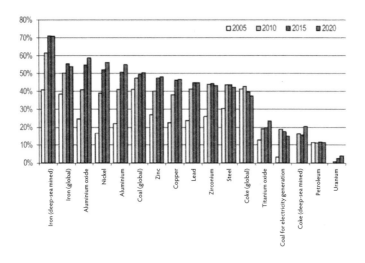

SOURCES: WM, World Bureau of Metal Statistics (WBMS), IHS, Platts, Bloomberg, Morgan Stanley Research, Business Insider, 2017.

Appendix 13

Main industrial applications of rare minerals

Resource	Applications
Antimony	Fire retardants (additive in plastics), polyethylene terephthalate catalyst
Baryte	Drilling mud for oil and gas drilling, glass industry, radioprotection, healthcare, metallurgy, pyrotechnics
Beryllium	Telecommunications and electronics, aerospace industry, civil and military nuclear power
Bismuth	Thermoelectric generators (automobiles), high-temperature superconductors, lead-free solder
Borate	Glass and ceramics
Cobalt	Mobile phones, computers, hybrid vehicles, magnets
Coking coal	Steel production
Fluorspar	Hydrofluoric acid, steel and aluminium production, ceramics, optics
Gallium	Semiconductors, light-emitting diodes (LEDs)
Germanium	Photovoltaic cells, fibre optics, catalysis, infrared optics
Indium	Computer chips, LCD screens
Magnesium	Aluminium alloys
Natural graphite	Electric vehicles, aerospace, nuclear industry
Niobium	Satellites, electric vehicles, nuclear industry, jewellery
Silicon metal	Integrated circuits, photovoltaic cells, electric isolators
Tantalum	Miniaturised condenser, superalloys
Tungsten	Cutting tools, shielding, electricity, electronics
Vanadium	Specialty steels, aerospace industry, catalysis

PGMs (platinum-group metals: ruthenium, rhodium, palladium, osmium, iridium, platinum)	Catalysts, jewellery
Rare earths (see table in the following appendix)	Permanent magnets, electric vehicles, wind turbines, TGV (high-speed train), medical scanners, lasers, fibre-optics data transmission, phosphors for plasma screens, security inks for banknotes, catalysis

SOURCE: French Parliamentary Office for Science and Technology Assessment (OPECST), French Geological Survey (BRGM), *Connaissance des énergies*, *Futura-Sciences*, Niobec, Lenntech.

Appendix 14

Main industrial applications of rare earths

Resource	Applications
Lanthanum	Superconductive compounds, lenses, lighting
Cerium	Catalytic converters, oil refinery, metal alloys
Praseodymium	Lighter flint, colourants, magnets
Neodymium	Permanent magnets, autocatalysts, oil refinery, lasers
Promethium	Luminescent compounds
Samarium	Magnets for missiles, permanent magnets, electric motors
Europium	Lasers, nuclear reactors, lighting, geo-chemistry, red phosphors in cathode-ray tubes
Gadolinium	Phosphorescent substance in cathode-ray tubes
Terbium	Green phosphor activator for cathode-ray tubes, permanent magnets
Dysprosium	Permanent magnets, hybrid engines
Holmium	Lasers, magnetism, superconductive compounds
Erbium	Long-distance fibre-optic communication, nuclear medicine
Thulium	Portable radiography, high-temperature superconductors
Ytterbium	Stainless steels, active ion (crystal lasers), portable radiography
Lutetium	Beta emitter (radiation)
Scandium	Lighting, marker, aluminium alloys
Yttrium	Red phosphors in cathode-ray tubes, superconductor alloys, fire bricks, fuel cells, magnets

SOURCE: French Senate, British Geological Survey, Economic Warfare School (EGE), Congressional Research Service, Portail de l'Intelligence Economique.

Appendix 15

Number of patents per country per year *

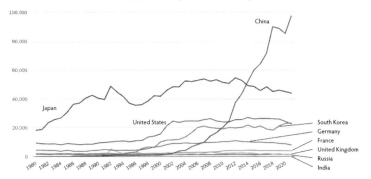

* Technologies: semiconductors, materials, surface technology, and coating.

SOURCE: World Intellectual Property Organization (WIPO), 2023.

Appendix 16

Geographic distribution of the EV value chain

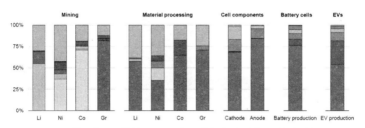

Notes: Li = lithium; Ni = nickel; Co = cobalt; Gr = graphite; DRC = Democratic Republic of Congo.

SOURCE: Global Supply Chains of EV Batteries, International Energy Agency (IEA), 2022.

Appendix 17

Rare metals used in a fighter aircraft

SOURCE: 'Strategic raw materials for defence. Mapping European industry needs', The Hague Centre for Strategic Studies, January 2023.

Appendix 18

Lifespan of the viable reserves
of the principal metals needed for the energy transition

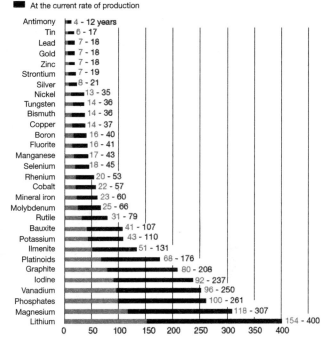

SOURCE: Table by L. Pennec for *L'Usine Nouvelle*, 2017.

Appendix 19

Rare-earth mines and deposits

● Mines

• Deposits

1 Bokan Mountain, USA
2 Aley, Canada
3 Rock Canyon Creek, Canada
4 Thor Lake, Canada
5 Niukluluk, Canada
6 Hardas Lake, Canada
7 Elliot Lake, Canada
8 Saint-Honoré, Canada
9 Strange Lake, Canada
10 Snowbird, USA
11 North Park, USA
12 Lemhi Pass, USA
13 Bald Mountain, USA
14 Bear Lodge, USA
15 Iron Hill, USA
16 Wet Mountains, USA
17 Pea Ridge, USA
18 Hicks Dome, USA
19 Carolina placers, USA

20 Green Cove Springs, USA
21 Mountain Pass, USA
22 Gallinas Mountains, USA
23 Pajarito Mountain, USA
24 Sierra de Tamaulipas, Mexico
25 Morro dos Seis Lagos, Brazil
26 Catalão I, Brazil
27 Araxá, Brazil
28 Serra Verde, Brazil
29 Poços de Caldas, Brazil
30 Serra do Itapirapuã, Brazil
31 Sarfartôq, Greenland
32 Motzfeldt, Greenland
33 Ilímaussaq, Greenland
34 Fen, Norway
35 Bastnäs, Sweden
36 Norra Kärr, Sweden
37 Olserum, Sweden
38 Sokli, Finland

39 Lovozero and Khibina complexes, Russia
40 Tomtor, Russia
41 Gatineau Dzora, Russia
42 Kharma, Russia
43 Ditrău, Romania
44 Kizilcaören, Turkey
45 Aksu Diamas, Turkey
46 Tamazeght complex, Morocco
47 Bou Naga, Mauritania
48 Nile Delta and Rosetta, Egypt
49 Kavonge (Gakara), Burundi
50 Mrima, Kenya
51 Wigu Hill, Tanzania
52 Kangankunde, Malawi
53 Songwe Hill, Malawi
54 Congolone, Mozambique
55 Ambohimirahavavy, Madagascar
56 Mandoro, Madagascar
57 Etaneno, Namibia

58 Okorusu, Namibia
59 Eureka, Namibia
60 Kalkfeld, Namibia
61 Ondurakorume, Namibia
62 Steenkampskraal, South Africa
63 Zandkopsdrift, South Africa
64 Pilanesberg Alkaline Complex, South Africa
65 Naboomspruit, South Africa
66 Phalaborwa (Palabora), South Africa
67 Rothards Bay, South Africa
68 Chavara, India
69 Amba Dongar, India
70 Samu, India
71 Orissa, India
72 Wajrakarur, China
73 Mushgai Khudag, Mongolia
74 Lugin Gol, Mongolia
75 Bayan Obo, China
76 Weishan, China

77 Maoniuping/Daluxiao, China
78 Mianya, China
79 Xunwu/Longnan, China
80 Nam Xe, Vietnam
81 Dong Pao, Vietnam
82 Northern Myanmar
83 Thai Peninsula, Thailand
84 Perak, Malaysia
85 Eneabba, Australia
86 Jangardup, Australia
87 Mount Weld, Australia
88 Brockman, Australia
89 Browns Range, Australia
90 Nolans Bore, Australia
91 Mary Kathleen, Australia
92 Olympic Dam, Australia
93 WIM 150, Australia
94 Dubbo Zirconia, Australia
95 Fraser Island, Australia

SOURCE: 'Global rare earth element (REE) mines, deposits and occurrences', British Geological Survey, May 2021.

Appendix 20

The seven pillars of the circular economy

SOURCE: 10 Key Indicators for Monitoring the Circular Economy, DataLab, French Ministry Of The Environment, Energy And Marine Affairs, In Charge Of International Relations On Climate Change, March 2017.

Appendix 21

France's mining potential

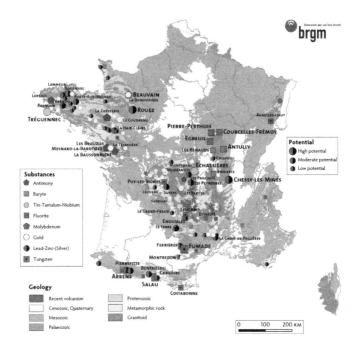

Appendix 22

Lithium mining projects in Europe

SOURCE: *Les Échos*, 27 September 2022.

Appendix 23

Map of deep-sea rare-metal resources

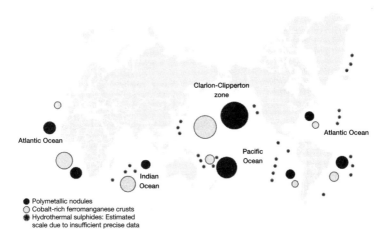

SOURCE: Atlante.fr, The International Seabed Authority, United Nations, French Geological Survey, 2022.

Notes

Preface

1 'Almost 400 new mines needed to meet future EV battery demand, data finds', *Tech Informed*, 27 September 2022.
2 Interview with Olivier Vidal, 2023.
3 The French government will need to invest €300 billion between now and 2030 to achieve the energy transition, as calculated in the report 'The economic implications of climate action', by Jean Pisani-Ferry and Selma Mahfouz, *France Stratégie*, November 2023.
4 See the consulting firm Deloitte's annual Circularity Gap Report. The 2023 report states that the global economy was only 7.2 per cent circular versus 9.1 per cent in 2018.
5 'Former industry CEO: 'For ecological transition, we must "exploit all of Europe's mines"', *Euractiv*, 11 January 2023.
6 See the SMD Mining Projects brochure, available on the SMD website (smd.co.uk).
7 'Bentley asking European Commission to raise weight limit for driver's license in EU', motori.com, 22 May 2022.

Introduction

1 Moderate consumption that thinker and writer Pierre Rabhi refers to as 'sobriety'. Refer to his book *The Power of Restraint,* Actes Sud, 2018.
2 Yuval Noah Harari, *Sapiens: a brief history of humankind*, HarperCollins, 2015.
3 United Nations Climate Change, 'Historic Paris Agreement on Climate Change: 195 Nations Set Path to Keep Temperature Rise Well Below 2 Degrees Celsius', 13 December 2015.

4   Distilling a tonne of orange blossom petals only produces a kilogram of
     essential oil.
5   It takes 500 kilograms of coca leaves to produce a kilogram of cocaine.
6   On average, a kilogram of rock contains 120 milligrams of vanadium, 66.5
     milligrams of cerium, 19 milligrams of gallium, and 0.8 milligrams of lutetium.
7   He and his colleagues, Jacob A. Marinsky and Lawrence E. Glendenin,
     produced it at the Oak Ridge National Laboratory in 1945.
8   Jeremy Rifkin, *The Third Industrial Revolution: how lateral power is transforming
     energy, the economy, and the world*, Palgrave Macmillan, 2011.
9   Jeremy Rifkin developed a number of 'master plans' for several cities
     and regions, including San Antonio in Texas, Monaco, Rome, Utrecht,
     and Luxembourg. The results are mixed — contested even. Read Erwan
     Seznec, 'Hauts-de-France: mais où est donc passée la troisième révolution
     industrielle?' ['Hauts-de-France: what happened to the third industrial
     revolution?'], Mediacités, 4 December 2020. In 2022, some 1,200 projects
     were run in the Hauts-de-France Region (northern France) by businesses,
     local authorities, training and research institutions, and associations in the
     construction, energy, mobility, and industrial or agricultural production
     sectors. The scope was extended to include issues pertaining to resource
     availability, biodiversity, conservation, and health impacts. See Claude
     Lenglet et al., 'Rev3. Une prospective 2022–2032', CCI Hauts-de-France,
     October 2022. In Luxembourg, the results are also mixed. Nevertheless, at
     the end of 2021, the minister for the economy extended the programme to
     meet the climate goals set by the 2015 Paris Agreement. Read the 'Third
     Industrial Revolution (TIR)' strategic study (2016), as well as the 2021 Status
     Report on the 49 Strategic Measures prepared by the Luxembourg Chamber
     of Commerce, and IMS (Inspiring More Sustainability) Luxembourg. Since
     December 2022, Jeremy Rifkin has been an advisor to the main opposition
     party in Turkey. Read 'CHP leader appoints US economist as advisor',
     Hürriyet Daily News, 1 December 2022.
10   Renewable energy includes numerous types of energy: hydraulic energy,
     biofuels, biomass, solar, and wind energy. The 15 per cent also includes
     renewable heat such as geothermal and solar thermal energy. Traditional
     biomass (wood energy, solids from agricultural waste) makes up a large
     proportion of renewable energy (around a quarter in 2020). See 'World Energy
     Outlook 2022', International Energy Agency, 2022.
11   The REPowerEU plan of May 2022 sets a new, highly ambitious target of 45
     per cent. See 'Factsheet: REPowerEU actions', European Commission, 18 May
     2022. Traditional biomass accounts for almost 60 per cent of renewable energy
     in the European Union. See 'Brief on biomass for energy in the European
     Union', Joint Research Centre, European Commission, 2019.
12   Christine Parthemore and John Nagl, 'Fueling the Future Force: preparing the
     Department of Defense for a post-petroleum era', Center for a New American
     Security, September 2010.
13   By 2030, the US army wants to have switched to low-carbon electricity at all

its military bases, including overseas, install a renewable energy microgrid at each base by 2035, and replace its non-tactical vehicles with electric vehicles. See *United States Army Climate Strategy*, Department of the Army, Office of the Assistant Secretary of the Army for Installations, Energy and Environment, February 2022. More in-depth analysis of the military's energy approach can be found in Constantine Samaras et al, 'Energy and the military: Convergence of security, economic, and environmental decision-making', *Energy Strategy Reviews*, vol. 26, November 2019.

14   Fuel and water supply convoys also pose a threat on the ground. According to a report by the Army Environmental Policy Institute, they accounted for over 3,000 deaths and injuries, or 10 to 12 per cent of US army casualties in Iraq and Afghanistan. Read David Eady et al., 'Sustain the Mission Project: casualty factors for fuel and water resupply convoys', Army Environmental Policy Institute, September 2009.

15   The French Innovation Agency has multiple high-technology telecommunications and robotics projects in progress. See the 2022 Defence Innovation Orientation Directive report ('DrOID 2022 Reference document: Guiding defence innovation'), as well as the 'Bilan d'activités 2021' ['2021 Activity report'] of the Ministry of the Armed Forces.

16   'On the front lines of Ukraine war: tech and social media', *South China Morning Post*, 18 February 2023.

17   The International Energy Agency's most optimistic fossil fuel reduction scenario would see global demand for oil fall by 76 per cent by 2050, that is, by 23 million barrels per day by 2050 (compared with 94.5 million barrels per day in 2021). Demand for natural gas would fall from 4,210 to 1,160 million cubic metres, and for coal from 5,640 to 540 million tonnes. See 'World Energy Outlook 2022', International Energy Agency, 2022.

18   The International Renewable Energy Agency's '1.5°C' scenario predicts the creation of 38 million jobs worldwide in renewable energy sectors by 2030, and more than 50 million jobs in other sectors linked to the energy transition: 'World Energy Transitions Outlook 2022: 1.5 Pathway', International Renewable Energy Agency (IRENA), March 2022.

19   See the 'white paper' by Florentin Krause, Hartmut Bossel, and Karl-Friedrich Müller-Reißmann, *Energie-Wende: Wachstum und Wohlstand ohne Erdöl und Uran* [*Energy Transition: growth and wellbeing without crude oil and uranium*], S. Fischer Verlag, 1980.

20   The 196 delegations comprised 195 states and the European Union.

21   At no point does the Paris Agreement on climate change include the words 'metals', 'minerals', or 'commodities'. Similarly, none of the decisions made at the COP 24 in Katowice (Poland) in December 2018 addresses mineral resources. As stated by the press service of the United Nations Framework Convention on Climate Change, 'we are not aware of a specific discussion on the question of mineral resources'.

22   Most rare metals cannot be substituted. Refer to the European Commission's 'Study on the Critical Raw Materials for the EU, 2023', as well as the EI/SR

substitution indices in the third list of critical raw materials for the EU, published in 2017.

23   'America and China are preparing for a war over Taiwan', *The Economist*, 9 March 2023. See also 'T-Day: The battle for Taiwan', *Reuters Investigates*, 5 November 2021.

24   'World Population Prospects 2022', Department of Economic and Social Affairs, Population Division, United Nations, 2022.

25   Bill Laws, *Fifty Plants that Changed the Course of History*, Firefly Books, 2010.

Chapter One: The rare metals curse

1   Gallium, for example, is a by-product of aluminium mining. Selenium and tellurium are associated with copper. Indium and germanium are zinc by-products. Philippe Bihouix and Benoît de Guillebon, *Quel futur pour les métaux? Raréfaction des métaux: un nouveau défi pour la société* [*What Does the Future Hold for Metals? Rarefication of Metals: a new challenge for society*], EDP Sciences, 2010, p. 33.

2   Updated in February 2022, the USGS list includes fifty minerals such as gallium, niobium, cerium, chrome, and lithium. (See 'US Geological Survey releases 2022 list of critical minerals', USGS.) The European Commission's inventory includes thirty-four critical raw materials: antimony, arsenic, aluminium/bauxite, baryte, beryllium, bismuth, borate, cobalt, coking coal, copper, feldspar, fluorspar, gallium, germanium, hafnium, helium, magnesium, manganese, natural graphite, nickel, niobium, phosphate rock, phosphorus, scandium, silicon metal, strontium, tantalum, titanium metal, tungsten, vanadium, platinum group metals, and heavy rare-earth elements, and light rare-earth elements. Some of the metals listed, such as silicon, are not considered 'rare' by geologists. The European Commission nevertheless qualifies them as 'critical' due to the threat to their supply. Shortages can often result from a lack in mining and refining infrastructure, adding the notion of industrial rarity to that of geological rarity. The scientific community uses the term 'rare metals' in reference to these two criteria. In addition, in 2023 the European Commission compiled a list of sixteen 'strategic' raw materials alongside its list of thirty-four 'critical' raw materials, based on their importance for the 'green transition' and digital technology (or defence and space). See 'Study on the Critical Raw Materials for the EU 2023' (final report), European Commission, 2023.

3   'Abundance of elements in the earth's crust and in the sea', *CRC Handbook of Chemistry and Physics*, 97th edition (2016–2017), pp. 14–17.

4   See 'Mineral Commodity Summaries 2023', United States Geological Survey (USGS), January 2023.

5   Ibid., pp. 72, 78, and 98.

6   Refer to the criteria set out in Report 782, *Key Issues with Strategic Metals: the case of rare earths* (submitted 23 August 2011) by Claude Birraux and Christian Kert, deputies of the French Parliamentary Office for Science and Technology

Assessment (OPECST). For a US critical mineral classification methodology, refer to the Draft Critical Mineral List — Summary of Methodology and Background Information — US Geological Survey Technical Input Document in Response to Secretarial Order No. 3359, Open-File Report 2018-1021, US Department of the Interior, US Geological Survey.

7   Harari, *Sapiens*, op. cit.

8   Such as praseodymium and neodymium.

9   'Travel to the Largest and Most Powerful MRI Magnet in the World', The French Alternative Energies and Atomic Energy Commission (CEA), 2 May 2017.

10   'Cinq questions sur les aimants, ces composants clés de la transition énergétique' ['Five questions on magnets, the key components of the energy transition'] , *Les Échos*, 11 January 2023.

11   These super magnets are produced with the rare-earth minerals neodymium and samarium that are alloyed with other metals, such as iron, boron, and cobalt. Magnets are usually 30 per cent neodymium and 35 per cent samarium. For the sake of clarity, they are more commonly referred to as 'rare-earth magnets' by the scientific community. Most electric motors do not need to be very powerful (for such applications as heating, ventilation, air conditioning, and lifts). They are mainly induction motors whose rotor contains an electromagnet (squirrel-cage rotor). Its magnetisation is not permanent; it is induced by the magnetic field of the stator, therefore not requiring a rare-earth magnet. But when more power is needed, electric motors or generators with rare-earth permanent magnets are indispensable (electric vehicles, wind turbines). The capacity of a 100 gram magnet containing neodymium is equivalent to that of a conventional 1 kilogram magnet without neodymium. See 'Mineral resources: Rare earths', French Geological Survey (BRGM), 12 July 2022.

12   Most modern electric vehicle motors contain rare-metal magnets, but carmakers are looking for alternatives to reduce their dependence on external, chiefly Chinese, suppliers. Motors without permanent magnets — and therefore without rare earths — include Renault's Zoé and Mégane E-Tech Electric (synchronous motor with wound rotor). Tesla has released a range of induction-motor vehicles. Other carmakers, such as BMW, Toyota, Volvo, and General Motors are following suit. Bear in mind, however, that these motors are slightly less efficient than motors with rare-earth magnets, particularly at low and moderate speeds (80–88 per cent compared with 90–95 per cent). They are generally less powerful, or bulkier and heavier. Tesla announced in 2023 the creation of a permanent-magnet motor without rare earths, but without specifying what its future components would be. See 'Pourquoi les moteurs électriques sans terres rares sont-ils l'avenir' ['Why rare-earth-free electric motors are the future'], *Auto Plus*, 22 March 2023.

13   See 'Critical Materials For The Energy Transition: rare earth elements, Technical Paper 2/2022', International Renewable Energy Agency (IRENA), May 2022. Offshore wind turbines use direct-coupled generators, which are

more reliable and require less maintenance. The rotor rotates at low speed and requires very powerful permanent magnets, which means rare earths. Around 500 kilograms of permanent magnets — one-third of which (neodymium) are rare earths — are needed for 1 MW of turbine power.

14 Especially solar panels made from copper indium gallium selenide (CIGS) solar cells.

15 A report by the European Commission in 2023 forecasts demand for 2050. A scenario of strong growth in non-fossil fuels and advanced technologies suggests a steep rise in demand for materials. Demand for neodymium (for permanent magnets in particular) could rise from 7 kilotonnes in 2020 to almost 30 kilotonnes in 2030, and 77 kilotonnes in 2050. Demand for gallium in photovoltaic applications is expected to rise to 49 tonnes in 2030 (from 23 tonnes in 2020) and could climb as high as 1,140 tonnes in 2050. See S. Carrara et al., 'Supply chain analysis and material demand forecast in strategic technologies and sectors in the EU — a foresight study', European Commission, 16 March 2023.

16 Light-emitting diode (LED) bulbs.

17 'Renewable Energy and Jobs. Annual Review 2022', International Renewable Energy Agency (IRENA), September 2022.

18 'Stepping up Europe's 2030 climate ambition. Investing in a climate-neutral future for the benefit of our people', Communication from the Commission to the European Parliament, the Council, the European Economic and Social Committee and the Committee of the Regions, COM/2020/562 final, European Commission, September 2020.

19 Refer to négaWatt's latest English-language report 'The energy transition at the heart of a societal transition: Overview of the 2022 négaWatt scenario', Association négaWatt, October 2021.

20 Gold, copper, lead, silver, tin, mercury, and iron.

21 British Petroleum 2017: Outlook for 2035. 'Gas is the fastest growing fuel (1.6 per cent p.a.); Oil continues to grow (0.7 per cent p.a.), although its pace of growth is expected to slow gradually; The growth of coal is projected to decline sharply: 0.2 per cent p.a. compared with 2.7 per cent p.a. over the past 20 years — coal consumption is expected to peak in the mid-2020s; Renewable energy is the fastest growing source of energy (7.1 per cent p.a.), with its share in primary energy increasing to 10 per cent by 2035, up from 3 per cent in 2015' (p. 15).

22 'Quand le monde manquera de métaux' ['When the World Runs Out of Metals'], Basta Mag, 26 September 2012.

23 See The Role of Critical Minerals in Clean Energy Transitions, IEA, 2021. For forecasts relative to Europe: S. Carrara et al., 'Supply chain analysis and material demand forecast in strategic technologies and sectors in the EU: a foresight study', European Commission, 16 March 2023.

24 Read Frank Marscheider-Weidemann, Sabine Langkau, Torsten Hummen, Lorenz Erdmann, and Luis Tercero Espinoza, 'Raw Materials for Emerging Technologies 2016', German Mineral Resources Agency (DERA), Federal Institute for Geosciences and Natural Resources (BGR), March 2016.

25   Roughly speaking, light rare-earth deposits are found in the north and
      west (Bayan Obo in Inner Mongolia, by far the largest deposit, Mianning in
      Sichuan province, and Shandong, not especially in the north or west). Heavy
      rare-earth deposits are found in the south and south-east (particularly in
      Jiangxi and Guangdong provinces). See J. Bai et al., 'Evaluation of resource and
      environmental carrying capacity in rare-earth mining areas in China', *Scientific
      Reports (Nature)*, 12, 6105, April 2022.

26   Surface area estimated by analysing satellite images, covering all types and sizes
      of mining operations. The total surface area worldwide is estimated at 100,000
      km². Six countries account for 52 per cent of the total: the Russian Federation,
      China, Australia, the United States, Indonesia, and Brazil. See V. Maus et al.,
      'An update on global mining land use', *Scientific Data*, 9, 433, July 2022.

27   'Environmental Disaster Strains China's Social Fabric', *The Financial Times*,
      26 January 2006.

28   'Toxic Mine Spill Was Only Latest in Long History of Chinese Pollution',
      The Guardian, 14 April 2011.

29   Further reading: 'Critical Materials For The Energy Transition: rare earth
      elements', Technical Paper 2/2022, IRENA, May 2022, (from p. 29), and P.
      Zapp et al., 'Environmental impacts of rare earth production', MRS [Material
      Research Society] Bulletin, 47, 2022.

30   R. Ganguli and D.R. Cook, 'Rare earths: a review of the landscape', *MRS Energy
      & Sustainability*, 5, 2018.

31   Accounting for 30 per cent of the world's known reserves, the Bayan Obo site
      in the Baotou district is the largest deposit in the world. Global production of
      rare earths in 2021 was estimated at 290,000 tonnes, including 168,000 tonnes
      from China, with the Bayan Obo site accounting for most Chinese production
      (around 70 per cent of light rare-earth production). See 'Mineral Commodity
      Summaries 2023', United States Geological Survey, January 2023, and 'Critical
      Materials For The Energy Transition: Rare Earth elements, Technical Paper
      2/2022', IRENA, May 2022.

32   One study estimates that over fifty years of mining (1958 to 2004) nearly 10
      million tonnes of mine tailings were stored in the open-pit tailings dams at
      Bayan Obo. See B. Li et al., 'In-situ gamma-ray survey of rare-earth tailings
      dams. A case study in Baotou and Bayan Obo Districts, China', *Journal
      of Environmental Radioactivity*, vol. 151, part. 1, January 2016. A more recent
      article mentions these regular spills into the Yellow River and neighbouring
      agricultural soils which were noted in 2010 by the local authorities. See C.
      Gramling, 'Rare earth mining may be key to our renewable energy future. But
      at what cost?', *Science News*, January 2023.

33   Y. Pan et H. Li, 'Investigating heavy metal pollution in mining brownfield and
      its policy implications: A case study of the Bayan Obo rare earth mine, Inner
      Mongolia, China', *Environmental Management*, vol. 57, 20 January 2016.

34   Some of Gao Xia's testimony features in the documentary *The dark side of green
      energies*, Grand Angle Productions, 2020, which I co-directed with Jean-Louis
      Perez.

35   The first opium war pitched China against the United Kingdom from 1839 to 1842, and was led by France, the United Kingdom, Russia, and the United States. The second opium war would last from 1856 to 1860.

36   The German concessions in Shandong, a province in the north of China, were handed over to Japan.

37   Founded by Sun Yat-sen, the Kuomintang party was defeated by the communist regime in 1949.

38   See Philippe Chalmin (dir.), *Des ressources et des hommes* ['Resources and people'], Nouvelles Éditions François Bourin, 2016.

39   See 'Study on the Critical Raw Materials for the EU 2023', European Commission, 2023, p. 23.

40   Interview with consultant Bruno Gensburger, Mutandis consultancy, 2016.

41   See 'CO2 Emissions in 2022', International Energy Agency, March 2023. These refer to CO_2 emissions, and not all greenhouse gases (methane, nitrous oxide, etc.). More detail on the emissions of EU members can be found on the database of the European Union's statistical office, Eurostat (https://ec.europa.eu/eurostat).

42   See E. Shifaw, 'Review of heavy metals pollution in China in agricultural and urban soils', *Journal of Health and Pollution*, vol. 18, June 2018.

43   This is according to a study by the Chinese Ministry of Water Resources published in 2016 (in Chinese). See 'China pollution: Over 80% of rural water in north-east "undrinkable"', *BBC News*, 12 April 2016.

44   See 'Is air quality in China a social problem?', *ChinaPower*, Center for Strategic and International Studies, updated in February 2021.

45   See 'Mineral Commodity Summaries 2023', United States Geological Survey, January 2023.

46   An estimated 15 to 25 per cent of cobalt production in the DRC originates from artisanal mining. See 'En RDC, explosion du cobalt artisanal, face noire de l'énergie verte' ['Explosion of artisanal cobalt mining in the DRC: the dark side of green energy'], *Le Monde*, 21 June 2023.

47   Exposure to cobalt (by inhaling or ingesting dust) can affect the heart, lungs, blood, and thyroid. Manganese, the observed concentrations of which are abnormal, is a neurotoxin. See C. Banza Lubaba Nkulu et al., 'High human exposure to cobalt and other metals in Katanga, a mining area of the Democratic Republic of Congo', *Environmental Research*, vol. 109, no. 6, August 2009, and C. Banza Lubaba Nkulu et al., 'Sustainability of artisanal mining of cobalt in DR Congo', *Nature Sustainability*, vol. 1, September 2018.

48   See 'Study on the Critical Raw Materials for the EU 2023', European Commission, 2023. Since 2017, the European Commission does not consider chrome a critical metal.

49   'Central Asia's longest river pollution dangerously high', *World Bulletin*, 2015.

50   See A. K. Kairakbaev et al., 'Environmental assessment of the quality of the Elek and Or rivers in the zone of influence of enrichment waste and metallurgical production', *IOP [Institute of physics] Conference Series: Materials Science and Engineering*, vol. 919 (062049), 2020.

51   The plateau region of the Andes Mountains holds 59 per cent of the world's known reserves, with Argentina alone accounting for 21 per cent. See C. Nugent, 'New lithium mining technology could give Argentina a sustainable gold rush', *Time*, 26 July 2022.

52   See 'Mining and its social impact in Latin America', *Deutsche Welle*, 16 September 2019, and 'Political changes boost risk for miners in Latin America', mining.com, 7 July 2022.

53   'Chile top court ratifies closure of Canadian-owned Pascua Lama mining project', *Reuters*, 14 July 2022.

54   F. Pearce, 'Why the rush to mine lithium could dry up the high Andes', *Yale Environment 360*, 19 September 2022.

55   'Rescue of Vale miners trapped underground in northern Ontario continues, with 35 of 39 now out', *CBC News*, 28 September 2021.

56   '10 years since South Africa's Marikana massacre', *Deutsche Welle*, 16 August 2022.

Chapter Two: The dark side of green and digital technologies

1   The 5th Annual Cleantech & Technology Metals Summit: Invest in the Cleantech Revolution.

2   'World Energy Outlook 2022', International Energy Agency, October 2022.

3   Tourillon reached this estimate using the Greenhouse Gas Equivalencies Calculator of the United States Environmental Protection Agency. Refer to the EPA's Greenhouse Gas Equivalencies Calculator online.

4   The sun's rays generate thermal energy, which heats fluids such as water that can then be used directly (using a solar water heater) or indirectly (steam passing through a generator to produce electricity).

5   The European Commission is funding applied research to reduce this water consumption, which far exceeds that of gas-fired power stations (1,000 litres/MWh). The research projects aim to reduce water consumption by 90 per cent. See 'Water consumption solution for efficient concentrated solar power', European Commission, 28 June 2019. See also Shabah Rohani et al., 'Optimization of water management plans for CSP plants through simulation of water consumption and cost of treatment based on operational data', *Solar Energy*, vol. 223, 15 July 2021.

6   Kimberly Aguirre, Luke Eisenhardt, Christian Lim, Brittany Nelson, Alex Norring, Peter Slowik, and Nancy Tu, 'Lifecycle Analysis Comparison of a Battery Electric Vehicle and a Conventional Gasoline Vehicle', UCLA Institute of the Environment and Sustainability, June 2012. For more information on the environmental impacts of electric batteries, see J. Sullivan and L. Gaines, 'A Review of Battery Life-cycle Analysis: state of knowledge and critical needs', Argonne National Laboratory, 1 October 2010.

7   According to some experts, it also costs at least a third of the total price of the vehicle.

8   An electric vehicle battery is made up of electrochemical cells (on average, two-thirds of the total mass), assembled in several modules, as well as

electrical and thermal management systems. The mass is often expressed as a percentage of the cell mass, rather than the battery as a whole. See L. Mathieu and C. Mattea, 'From dirty oil to clean batteries: Batteries vs. oil: a systemic comparison of material requirements', *Transport & Environment*, March 2021, and 'Environmental challenges through the life cycle of battery electric vehicles', Policy Department for Structural and Cohesion Policies, Directorate-General for Internal Policies, PE 733.112, European Parliament, March 2023.

9   Kimberly Aguirre et al., 'Lifecycle analysis comparison of a battery electric vehicle and a conventional gasoline vehicle', op. cit. On the environmental impact of electric batteries, see also J. Sullivan, L. Gaines, 'A review of battery life-cycle analysis: State of knowledge and critical needs', Argonne National Laboratory (Illinois, USA), ANL/ESD/10-7, 1 October 2010.

10 Calculated on the basis of a compact saloon car with a 60-kWh battery. See *Les avis de l'Ademe. Voitures électriques et bornes de recharges ['Ademe insights on electric cars and charging stations]*, Ademe, October 2022.

11 See F. Euvrard, 'Tesla: La berline Model 3 Grand Autonomie RWD enfonce le clou mais ...' ['Tesla: The Model 3 Long Range AWD drives its point home, but ...'], *The Automobilist*, 20 April 2023.

12 Francesco Del Pero, Massimo Delogu, Marco Pierini, 'Life Cycle Assessment in the automotive sector: A comparative case study of Internal Combustion Engine (ICE) and electric cars', *Procedia Structural Integrity*, vol. 12, 2018, pp. 521–37.

13 John Petersen's articles are available on seekingalpha.com.

14 Rifkin, *The Third Industrial Revolution*, op. cit.

15 Rifkin, *The Zero Marginal Cost Society: the internet of things, the collaborative commons, and the eclipse of capitalism*, Palgrave Macmillan, 2014.

16 'US Car Sharing Service Kept 28 000 Private Cars Off the Road in 3 Years', *The Guardian*, 23 July 2016.

17 Eric Schmidt and Jared Cohen, *The New Digital Age: reshaping the future of people, nations and business*, Knopf, Random House Inc., 2013.

18 Imagine the staggering amounts of digital data this already generates. 'Every two days now we create as much information as we did from the dawn of civilization up until 2003,' say Schmidt and Cohen. This outlook has an economic impact: the turnover of the digital technologies sector (computer terminals, services, telecommunications, data centres, Internet of Things, robotics, etc.) stood at around $6,800 billion in 2023, and could reach $9,000 billion in 2027. See 'IT Services Worldwide', 'Technology', 'Market Insights' on the Statista website (statista.com), consulted in April 2023.

19 Every minute around the world, 2,400 trees are felled, which is the equivalent of a third of the land area of France every year. See *'Déforestation: 18 millions d'hectares de forêts perdus en 2014'* ['Deforestation: 18 million hectares of forest lost in 2014'], *Le Monde*, 3 September 2015. 'In the tropical domain, net annual loss of forest area from 2000 to 2010 was about 7 million hectares, and net annual increase in agricultural land area was more than 6 million hectares';

2016 State of the World's Forests, *Food and Agriculture Organisation of the United Nations*, 2016.

20   Consult the Raw Materials Information System of the European Commission for data on gold and silver and their use by economic sector.

21   Guillaume Pitron, *The Dark Cloud: how the digital world is costing the earth*. Translated by Bianca Jacobsohn. Scribe, 2023.

22   Fabrice Flipo, Michelle Dobré, and Marion Michot, *La Face cachée du numérique. L'impact environnemental des nouvelles technologies* [*The Dark Side of Digital: the environmental impact of new technologies*], L'Échappée, 2013.

23   Ibid, quoting Frédéric Bordage, *Sobriété numérique, les clés pour agir* ['Digital sobriety: the keys for action'], Buchet-Chastel, 2019.

24   Coline Tison and Laurent Lichtenstein, *Datacenter: The hidden pollution*, Documentary, Camicas Productions, 2013.

25   See *Analyse comparée des impacts environnementaux de la communication par voie électronique* ['Comparative study of the environmental impact of electronic communication'], French Environment & Energy Management Agency (ADEME), July 2011.

26   *Email Statistics Report, 2021-2025*, The Radicati Group (United States), 2021.

27   The documentary mentioned above takes us as far as the coalmines in Appalachia, in West Virginia, where the fossil fuel resources used to run the US power stations are extracted. 'There's nothing virtual about our clicks,' states the documentary. Referring to the illusion of dematerialisation, it poses the question: 'will our emails ultimately destroy the Appalachia mountains?'

28   See *Le numérique en Europe: une approche des impacts environnementaux par l'analyse du cycle de vie* ['Digital technologies in Europe: measuring the environmental impact using lifecycle analysis'], GreenIT and NegaOctet, December 2021. For more on the impact of aviation, which accounts for 5.2 per cent of Europe's greenhouse gas emissions, refer to the *European aviation environmental report 2022*, European Environment Agency/EASA (European Union Aviation Safety Agency), 2022.

29   As an example, in 1951, forty-four UNIVAC I computers (Universal Automatic Computer I) — the first US commercial computer — were sold. In 2021, over 1.4 billion smartphones and 340 million personal computers were sold worldwide. See the press releases of the Gartner consultancy firm: 'Gartner says global smartphone sales grew 6% in 2021', March 2022, and 'Gartner says worldwide PC shipments declined 28.5% in fourth quarter of 2022 and 16.2% for the year', January 2023.

30   'It's less about scarce resources and more about "full dustbins",' write experts PierreNoël Giraud and Timothée Ollivier in *Économie des matières premières* [*The Economy of Commodities*], La Découverte, coll. Repères, 2015.

31   See 'Déchets, chiffres-clés. Edition 2023' ['Waste, key figures. 2023 edition'], ADEME, 2023. For volumes in circulation, produced and recycled in the European Union, refer to 'Waste electrical and electronic equipment (WEEE) by waste management operations', Data Browser, *Eurostat*, March 2022.

32 See 'La consommation de métaux du numérique: un secteur loin d'être dématérialisé' [Digital's consumption of metals: a sector far from dematerialised], *France Stratégie*, June 2020.

33 'End-of-life recycling rates for selected metals', International Energy Agency (IEA), October 2022.

34 See 'Number of collected mobile and personal handy phone system (PHS) terminals for recycling purposes in Japan from fiscal year 2013 to 2020', *Statista*, August 2022, and Tunali, M., Tunali, M. M., & Yenigun, O., 'Characterization of different types of electronic waste: heavy metal, precious metal and rare earth element content by comparing different digestion methods', *Journal of Material Cycles and Waste Management*, 23, 2021. This estimate is based on the average weight of a smartphone (120g) and the rare earth concentrations given in the publication in g/kg.

35 Estimated volume. Refer to 'The Global E-Waste Monitor 2020', Unitar (United Nations Institute for Training and Research), *E-Waste Monitor*, 2020. Japan generated 2,500,000 tonnes of e-waste (p. 74), 20 per cent of which is digital waste (p. 24).

36 Cerium, for instance — a rare earth used to polish glass — can be replaced with zirconium.

37 New uses have made it possible to reduce the quantity of europium and terbium in fluorescent lamps by 80 per cent, and the quantity of dysprosium used in magnets by 30 per cent. European car manufacturers are even looking for ways to make magnets without the need for rare-earth metals.

38 Interview with Jack Lifton, co-founder of the Technology Metals Research and Critical Mineral Institute, 2016. The Davis-Monthan US Air Force base is the world's largest military aircraft scrapyard, with over 3,000 aircraft. Further reading: 'World's Largest Military Aircraft Boneyard: World record in Tucson, Arizona', *World Record Academy*, 16 March 2023.

39 'The US left billions worth of weapons in Afghanistan', *Foreign Policy*, 28 April 2022.

40 Surendra M. Gupta, *Reverse Supply Chains: issues and analysis*, CRC Press, 2013. Read also Rémy Le Moigne's fascinating 'L'Économie circulaire: comment la mettre en oeuvre dans l'entreprise grâce à la reverse supply chain?' [*Using the Reverse Supply Chain to Implement the Circular Economy in Business*], Dunod, 2014.

41 Metal prices were structurally low between end-2014 and 2019, before resuming an upward trend. See 'Average price of selected rare earth oxides from 2015 to 2022', *Statista*, January 2023, and 'Average lithium carbonate price from 2010 to 2022', *Statista*, January 2023.

42 See *Metals for clean energy: Pathways to solving Europe's raw materials challenge*, Eurometaux — KU Leuven (Université catholique de Louvain), April 2022. The following two papers provide extensive insights on certain metals: Matos, C. T., et al., *Material system analysis of nine raw materials: barytes, bismuth, hafnium, helium, natural rubber, phosphorus, scandium, tantalum and vanadium*, Joint Research Centre (JRC), The European Commission's Science and Knowledge Service, Publications Office of the European Union, 2021; and Matos, C. T.,

et al., *Material System Analysis of five battery-related raw materials: cobalt, lithium, manganese, natural graphite, nickel*, JRC, Publications Office of the European Union, 2020.

43   Interview with Christian Thomas, founder of Terra Nova Développement (TND), 2017.

44   Basel Convention on the Control of Transboundary Movements of Hazardous Wastes and Their Disposal, adopted in Basel on 22 March 1989.

45   In particular, the convention bans waste containing hexavalent chromium compounds, copper, zinc, cadmium, or antimony.

46   In 2018, members of the Basel Action Network conducted a study by fitting GPS beacons on over 300 items of waste collected in Europe to geolocate their route. Six per cent of this waste was exported to developing countries such as Nigeria or Ghana, most of it probably illegally. Read 'Holes in the circular economy. WEEE leakage from Europe. A report of the e-Trash Transparency Project', Basel Action Network, 2018.

47   *EU Serious and Organised Crime Threat Assessment (SOCTA)*, Europol, 2013. Illegal waste trafficking is still included in the environmental crime section in 2021. Since the Covid-19 pandemic, Europol believes that illegal waste management may even have increased. As a result, the agency has launched campaigns to inspect waste recyclers and transporters. *EU Serious and Organised Crime Threat Assessment (SOCTA)*, Europol, December 2021. It is difficult to put a figure on the quantity of e-waste. According to Eurostat's online database, 4.7 million tonnes were collected in 2020 in the European Union, out of the 12 to 13 million tonnes of waste generated. Based on the 6 per cent export figure estimated by Basel Action Network in 2018, exported waste could be as high as several hundred thousand tonnes.

48   Waste exports in 2020 outside the European Union stood at 32.7 million tonnes, of which 30 per cent are believed to be illegal. Read 'European Green Deal: Commission adopts new proposals to stop deforestation, innovate sustainable waste management and make soils healthy for people, nature and climate', press release of the European Commission, 17 November 2021. See also 'Waste shipments: MEPs push for tighter EU rules', press release of the European Parliament, 17 January 2023.

49   Martin Eugster and Roland Hischier, 'Key Environmental Impacts of the Chinese EEE-Industry', Tsinghua University, China, 2007.

50   See 'Les terres rares: des propriétés extraordinaires sur fond de guerre économique' ['Rare-earth Metals: extraordinary properties against a backdrop of economic war'] with Paul Caro, rare-earth metals expert at Canal Académies (Institut de France). See also 'China Tries to Clean Up Toxic Legacy of its Rare Earth Riches', *New York Times*, 22 October 2013.

51   'Rare Earth and Radioactive Waste: a preliminary waste stream assessment of the Lynas advanced materials plant, Gebeng, Malaysia', National Toxics Network, April 2012. Likewise, in Malaysia, where our investigation led us in 2011: between the end of the 1970s and 1994, the Japanese multinational corporation Mitsubishi mined and refined rare-earth metals in Bukit Merah, in

the north of the country. As explained to us by environmental activist Tan Ka Kheng, 'Their activities generated tremendously high levels of radiation. This in the nuclear industry is considered medium-activity radioactive waste and must therefore be manoeuvred with the greatest of care. But ... they disposed of the waste in old, rusty barrels and used plastic bags as cladding. What a tragedy! Mitsubishi then closed the facility and left! Now we're left with this waste for 14.4 billion years!' Bukit Merah is one of the most radioactive sites in Asia. While we were there, $100 million were committed to renovate the site. Watch the author's and Serge Turquier's documentary *Rare Earths: the dirty war*, 2012.

52   A 2020 report by the World Bank estimates that the production of aluminium, graphite and nickel for new energies could emit 1.4 billion tonnes of CO_2 equivalent by 2050 — equal to the 2018 emissions of France, Germany and the UK combined. While emissions from renewable energies represent only 6 per cent of coal and gas emissions, they are far from negligible. See 'Minerals for climate action: The mineral intensity of the clean energy transition', The World Bank Group, Mineral Choices, 11 May 2020.

53   'The Growing Role of Minerals and Metals for a Low Carbon Future', *World Bank Group*, June 2017.

54   '"Electrification Puts the Car Industry at Risk" Says PSA Boss Tavares at Frankfurt', *The Telegraph*, 12 September 2017. Carlos Tavares also stated that 'All this agitation and chaos will come back to haunt us because we will have made the wrong decisions that are insufficiently considered, lack perspective, and are made based on day-to-day emotions.' A year later, Yoshihiro Sawa, president of the luxury carmaker Lexus, warned against electric vehicles: 'EVs currently require a long charging time and batteries that have an environmental impact at manufacture and degrade as they get older. And then, when battery cells need replacing, we have to consider plans for future use and recycling. It's a much more complex issue than the current rhetoric perhaps suggests. I prefer to approach the future in a more honest way.' See 'Lexus Boss on EVs, Autonomy and Radical Design', *Autocar Professional (India)*, 11 August 2018.

Chapter Three: Delocalised pollution

1   For further reading, see Philippe Bihouix and Benoît de Guillebon, *Quel futur pour les métaux? Raréfaction des métaux: un nouveau défi pour la société* [*What Does the Future Hold for Metals? Rarefication of Metals: a new challenge for society*], op. cit.

2   See Mineral Resources Online Spatial Data, US Geological Survey.

3   'Les enjeux des métaux stratégiques: le cas des terres rares', parliamentary report no. 782 (French Senate), submitted on 23 August 2011.

4   Ibid.

5   See 'Rare Earth Mining at Mountain Pass', *Desert Report*, March 2011.

6   The United States owns and operates at least 2 per cent of the world's rare earth mines. See 'Mineral Commodity Summaries 2023', US Geological Survey, January 2023.

7   Interview with Chen Zhanheng, deputy secretary general of the Association of China Rare Earth Industry, 2016.
8   Ten years later, MP Materials bought the site and reopened the mine, focusing on cerium production, which was more profitable at the time. However, the mine went bankrupt again in 2015. Since then, the United States has opened its eyes to its dependence on China, and the government has released $45 million in aid for MP Materials, Molycorp's successor. MP Materials now produces 15 per cent of the world's rare earths, or 42,000 tonnes. However, the ores are still exported to China for refining. Massive investments have recently been made to reshore refining operations and open a rare earth magnet plant in the US. Read 'In the Mojave Desert, the rebirth of the only American rare earth mine', *Le Monde*, 2 January 2023.
9   Interview with Eric Noyrez, former CEO of Lynas and current managing director of Serra Verde Rare Earth, 2019.
10 'The global economic crisis put the project on hold for over a year. In September 2009, Nicholas Curtis managed to raise enough capital to revive the project, and engineering works began in January 2010. In the end, the Mount Weld mine produced its first gram of rare earths in 2013,' adds Noyrez.
11 Interview with Jean-Yves Dumousseau, the then sales director of US chemicals company Cytec, 2011.
12 Nicolas Hulot later became the environment minister under the Macron presidency.
13 See Régis Poisson, 'La guerre des terres rares' ['The Rare Earths War'], *L'Actualité chimique*, no. 369, December 2012.
14 Operations could resume: in September 2022, the Solvay group inaugurated a research unit and announced investments in La Rochelle to supply markets with rare earth permanent magnets. Read the news releases 'Solvay inaugurates state-of-the-art innovation pilot unit for solid state batteries in Europe', and 'Solvay to develop major hub for rare earth magnets in Europe', on Solvay's website, 16 September 2022.
15 Interviews with Jean-Paul Tognet, former industrial and raw materials director at Rhône-Poulenc and Rhodia Terres Rares, 2016 and 2017.
16 The La Rochelle site stores 20,000 tonnes of thorium hydroxide, 11,000 tonnes of thorium nitrate, and several thousand tonnes of residues produced until 1994 when monazite was processed to extract rare earths. This type of waste is no longer produced today. See 'Bilan de l'état radiologique de l'environnement français en 2010-2011' ['Assessment of the radiological status of the French environment in 2010-2011'], French Institute for Radiation Protection and Nuclear Safety (IRSN), 2011, and the inventory records of the French National agency for the management of radioactive waste (Andra): 'F6-8-01: Colis de résidus radifères RRA (Solvay)' ['Batches of radioactive waste (Solvay)'], and 'F6-8-07: Résidus de valorisation des hydroxydes bruts de thorium HBTh (Solvay)' ['Recovered residue of crude thorium hydroxide'], Andra, 2021.
17 The shore of Port Neuf lies opposite the Minimes marina, less than 2 kilometres from the Saint Nicolas tower.

18   The radioactive residue has probably dispersed in the ocean since then, as radioactivity levels have fallen below the critical limits. In 2009, the Radiation Protection and Nuclear Safety (IRSN) took readings in the port of La Rochelle and found no abnormal radioactivity. See 'Bilan de l'état radiologique de l'environnement français en 2010-2011', ['Assessment of the radiological status of the French environment in 2010-2011'], op. cit, p. 260. In 2019, the French Radiation Protection Society (SFRP) took new readings to estimate the impact of radioactivity on species living in the Bay of La Rochelle (algae, crustaceans, pelagic fish, etc.). The readings measured below the critical limits. Radioactivity is therefore considered to have no significant impact on local species. See 'Evaluation du risque radiologique environnemental de l'écosystème marin dans la baie de La Rochelle via l'outil ERICA' ['Assessment of the environmental radiological risk on the marine ecosystem using the ERICA tool'], SFRP presentation, 18 June 2019.

19   Alain Roger and François Guéry (dir.), *Maîtres et protecteurs de la nature* [*Masters and Protectors of Nature*], Champ Vallon, 1991. According to Régis Poisson, a former engineer at Rhône-Poulenc, one day the local authorities apparently asked: 'Can't you take the red out of the plumes to make them look less dirty?'

20   'La CRIIRAD crie à la radioactivité dans la baie de La Rochelle' ['CRIIRAD Exposes Radioactivity in the Bay of La Rochelle'], *Libération*, 19–20 March 1988.

21   The problem appears to have lasted until at least 2002, when a new report from the CRIIRAD demonstrated 'the ongoing contamination of the old outfall pipe' on the shore of Port Neuf. Refer to CRIIRAD report no. 10-149 VI 1, 'Mesures radiamétriques sur terrain de l'université de La Rochelle' ['Radiometric In-field Measurements by La Rochelle University'], 15 December 2010. According to Bruno Chareyron, 'The facility had not been dismantled and the immediate environment of the outfall pipe wasn't even cordoned off. The company had not done everything in its power to limit local residents' exposure to radiation.'

22   'La CRIIRAD crie à la radioactivité dans la baie de La Rochelle' ['CRIIRAD Exposes Radioactivity in the Bay of La Rochelle'], op. cit.

23   Interviews with Jean-Paul Tognet, 2016 and 2017.

24   Jean-Yves Le Déaut, 'Report on Low-level Radioactive Wastes', *Parliamentary Office for Science and Technology Assessment (OPECST)*, 1992.

25   Interviews with Jean-Paul Tognet, 2016 and 2017.

26   Ibid. I contacted Jean-René Fourtou, but he did not respond to my request for an interview.

27   Interview with Jean-Yves Dumousseau, 2016. Jean-Paul Tognet states that the Chinese prices are around 25 per cent lower than those of the competition.

28   Ibid.

29   'Toxic Memo', *Harvard Magazine*, 5 January 2001.

30   Interview with Patrice Christmann of BRGM, the French Geological Survey, 2013.

31   Regulation (EC) No 1907/2006 of the European Parliament and of the Council of 18 December 2006 concerning the Registration, Evaluation, Authorisation

and Restriction of Chemicals (REACH), establishing a European Chemicals Agency. The regulation should be revised in 2023 to make it more binding. See 'Pollution: La réglementation européenne sur les substances chimiques doit être révisée d'urgence' ['Pollution: European regulations on chemicals must be revised as a matter of urgency', an article by a collective of organisations specialising in environmental health', *Le Monde*, 8 April 2023.

32   The population of the European Union (27 member states since Brexit) stood at 447 million people at 1 January 2022. See 'EU population continues to decrease for a second year', *Eurostat*, 11 July 2022.

33   Interview with Christophe-Alexandre Paillard, then Director of the Armament and Defense Economy Department at the French Institute for Strategic Research Affairs Directorate (IRSEM), 2013.

34   As rightly put by Louis Maréchal, currently extractives sector policy adviser within the OECD's Responsible Business Conduct Unit, by transferring the responsibility of production to mining countries we have also relocated the associated societal repercussions: corruption, conflict, governance issues, the black market, human rights violations, and so on. Interview with Louis Maréchal, 2017. Dr Denis Mukwege, who was awarded the Nobel Peace Prize in 2018 for his work on behalf of women victims of violence, makes an unequivocal link between the extraction of strategic minerals and the endless conflicts ravaging his country. He calls on consumers to act: 'Turning a blind eye to this tragedy is being complicit'. See Denis Mukwege's profile on the website of The Nobel Prize for his acceptance speech made on 10 December 2018 in Oslo.

35   Keynote by Gregory Bowes, Chairman of Northern Graphite, 5th Annual Cleantech & Technology Metals Summit: Invest in The Cleantech Revolution, 2016.

36   See 'Mining and Metals: A power base for all nations. Focus of mining 1850-2030', *RMG Consulting*, Concord (Ontario, Canada), 2021.

37   Interview with Lan Xue, professor of political science at Tsinghua University, 2016.

38   See Tesla's 2022 Impact Report. The Model 3 and Model Y motors contain rare earths, the quantity of which Tesla has made significant efforts to reduce (down 25 per cent in 2023 compared with 2017). Read 'Elon Musk wants a future of EV motors without rare earth metals', *Popular Mechanics*, 2 June 2023.

39   To give just one example, 'The growing footprint of digitalisation', UN Environment Programme, November 2021.

40   'The Fairphone, a No-conflict Smartphone without Planned Obsolescence', *World Forum for a Responsible Economy*, 28 July 2017.

41   'The anti-waste law in the daily lives of French people: What does that mean in practice?', Reference Document, Ministry of Ecological Transition, Government plan for the Circular Economy, September 2021 (Law no. 2020-105 of 10 February 2020 on the fight against waste and the circular economy.

42   See the 'Circular economy action plan', which lists the measures adopted since 2015 (under Environment, then Strategy), on the website of the European Commission.

43   See 'Pacte vert et paquet climat: l'UE vise la neutralité carbone dès 2050'
       ['Green Deal and climate package: the EU is aiming for carbon neutrality by
       2050'], *Vie publique* (DILA — French Directorate of Legal and Administrative
       Information, Office of the Prime Minister), August 2021.

44   See 'By the numbers: The Inflation Reduction Act', The White House,
       15 August 2022.

45   This reality certainly helps to explain a discussion that 'fell on deaf ears'
       between Song Xi Chen, economics professor at Peking University, and a US
       steel trader from Milwaukee, both travelling on a plane from Chicago to
       Beijing in 2014. 'I said to him: "You are responsible for the pollution in China!"
       To which he replied: "But I don't own the industry!"' Interview with Song Xi
       Chen, 2016. Note, however, that in 2023 the European Parliament will adopt
       the 'border carbon adjustment mechanism', which aims to apply the criteria
       of the European carbon market to imports. This mechanism will be phased in
       gradually until 2034. See 'The debacle of Europe's first major carbon market,
       the Emissions Trading System', *Le Monde*, 11 June 2023.

46   Throughout the 1990s and up until 9/11, US military spending declined
       at an almost constant rate, from $570 billion (in 2020 dollars) in 1992 to
       $480 billion in 2000, down 20 per cent. From 2001, spending began to rise
       substantially, reaching a record $875 billion in 2010. It has since declined
       slightly, but is on a stable trend ($767 billion in 2021). In contrast, France's
       military spending fell by 10 per cent over the same period ($46 billion
       in 2000). While annual budgets have varied since 2001, there has been
       an increase since 2015, with spending totalling $54 billion in 2021. With
       a budget of 413 billion euros for the 2014–2030 period, France's military
       programming law gives this trend roots. See the Stockholm International
       Peace Research Institute's SIPRI Military Expenditure Database (2022
       data), and 'L'Assemblée nationale adopte largement la loi de programmation
       militaire' ['The National Assembly passes the military programming law by
       a large majority'], *Le Monde*, 8 June 2023.

47   Interview with Alain Liger, former secretary-general of the Comité pour les
       métaux stratégiques [COMES – Committee for Strategic Metals], 2016.

48   Namely, the National Defense Stockpile (NDS), created in 1939 and managed
       by the US Department of Defense. Strategic stocks related to healthcare and
       energy are managed by other federal agencies. In the 1980s, the NDS was worth
       $14.8 billion. Over time, a series of budgetary pressures led the US Congress to
       authorise the stockpile to be liquidated, and for the proceeds to be used to buy
       and maintain military equipment. By 2009, the NDS was reduced to a value
       of $1.4 billion. See K. H. Butts, B. Bankus, and A. Norri, 'Strategic minerals: Is
       China's consumption a threat to United States security?', Center for Strategic
       Leadership (CSL) — US Army War College, vol. 7–11, July 2011, and *Managing
       materials for a twenty-first century military*, National Academies Press, Washington,
       DC, 2008 (ch. 10). Today, the United States is reinvesting hundreds of millions
       of dollars to replenish its stocks. See 'Fact sheet: Securing a Made in America
       supply chain for critical minerals', The White House, 22 February 2022.

49   Interview with Jean-Philippe Roos, then commodity market analyst in the Natexis Asset Management economic research department, 2010. These transactions apparently began even earlier: after signing the treaties resulting from the Strategic Arms Limitation Talks (SALT) I and II, the USSR dismantled some of its atomic bombs and sold the uranium to the US. The massive quantity of minerals that suddenly flooded the market also contributed to 'killing' the US uranium industry.

50   'Braderie forestière au pays de Colbert' ['The Sell-off of the Forestry Industry in the Country of Colbert'], *Le Monde diplomatique*, October 2016, 'La Chine siphonne la forêt française' ['China is siphoning off France's forests'], *Les Echos*, 9 August 2021, and 'Trafic de bois : l'Etat ferme les yeux sur le pillage des forêts publiques' ['Timber trafficking: the government turns a blind eye to the plundering of public forests'], *Disclose*, 21 February 2023.

51   'Grasse se remet au parfum' ['Grasse Back on the Scent'], *M. Le magazine du Monde*, 11 July 2016.

52   Thales registration document, 2015.

53   Some industrial groups have recently adopted strategies that break with Toyotism. Elon Musk, for instance, insists on 'vertical integration' in all his companies, i.e. full control over all stages of the value chain, including downstream mining. See 'Does Elon Musk have a strategy?', *Harvard Business Review*, 15 July 2022.

54   In France, a Senate report talks of the 'unthinkable of the energy transition'. See 'Cinq plans pour reconstruire la souveraineté économique' ['Five plans to rebuild economic sovereignty'], report by the Economic Affairs Committee No. 755 (2021-2022), Senate, 6 July 2022.

55   Interview with Alain Liger, 2016.

56   In Portugal, the BRGM geologists discovered the Neves-Corvo copper orebody. In Quebec, their exploration activities enabled the mining of copper · and zinc orebodies, the Langlois mine.

57   Interview with Alain Liger, 2016.

58   Ibid.

59   'Trends in the Mining and Metals Industry', ICMM (International Council of Mining & Metals), October 2012.

60   Jean-Marie Guéhenno, French White Paper on Defence and National Security 2013, Ministry of Defence, 2013. Since 2017, White Papers have been replaced by National Strategic Reviews ('RNS'), which are more succinct and updated more regularly. The following two reviews focus on the concept of strategic self-reliance, but do not address the issues related to raw materials or rare metals. See 'Defence and National Security Strategic Review 2017', French Ministry of Armed Forces, and 'National Strategic Review', November 2022, General Secretariat for Defence and National Security (SGDSN), November 2022.

61   Upon his retirement in 1984, Alain de Marolles wrote a predictive analysis in which he referred to a 'third industrial revolution' in progress, spurred by the electronics industry, the space race biology. He commented on the massive

demand for metals generated by these sectors, and predicted that some of these metals — copper, cobalt, manganese, nickel, platinum, and gold — would in the future be mined from the bottom of the sea. See Alain de Gaigneron de Marolles, *L'Ultimatum: fin d'un monde ou fin du monde?* [*Ultimatum: end of one world or end of the world?*], Plon, 1984.

62   In 2022, an estimated 300,000 tonnes of rare earths were produced globally, and the number of people on the planet reached eight billion. See 'Mineral Commodity Summaries 2023', US Geological Survey, op. cit.

63   Interview with Jack Lifton, 2016.

64   Interview with Didier Julienne, natural resources strategist, 2016.

65   '50 Years Ago: cargo cults of Melanesia', *Scientific American*, 1 May 2009.

66   'The Surprising Number of American Adults Who Think Chocolate Milk Comes from Brown Cows', *The Washington Post*, 15 June 2017.

Chapter Four: The West under embargo

1   Including, of course, the 50 mineral commodities considered critical and strategic by the US Geological Survey in 2023.

2   2021 production volumes. See *Mineral Commodity Summaries*, US Department of the Interior, US Geological Survey, January 2023.

3   See Table A, p. 6 of 'Study on the critical raw materials for the EU 2023', European Commission, 2023.

4   See 'Increasingly strategic raw materials', Blog of Commissioner Thierry Breton, 12 June 2022.

5   'Study on the critical raw materials for the EU 2023', op. cit.

6   Interview with Felix Preston, then energy, environment and resources specialist at Chatham House (Royal Institute of International Affairs), 2016.

7   In 2019, China consumed precisely 3.09 billion tonnes of metals. See the statistics on the domestic consumption of metals by country on the OECD Statistics platform (oecd.stat) under Environment>Material resources>Domestic Material Consumption>Metals.

8   Geneviève Barman and Nicole Dulioust, 'Les années françaises de Deng Xiaoping' ['Deng Xiaoping's French Years'], *Vingtième Siècle, Revue d'histoire*, no. 20, October-December 1988, pp. 17–34.

9   'Le singe et la souveraineté des ressources' ['The Monkey and the Sovereignty of Resources'], *Le Cercle, Les Échos*, 12 February 2016. Wen Jiabao, Hu Jintao's Prime Minister from 2003 to 2013, trained as a geologist. Xi Jinping, the current President of the People's Republic of China, studied chemical engineering. Li Qiang, his Prime Minister since 2023, has an initial training in mechanised agriculture. Zhang Guoqing and Liu Guozhong, vice-premiers since 2023, have degrees in defence technology (electrical engineering and artillery systems). See 'Chinese technocrats 2.0: How technocrats differ between the Xi Era and Jiang-Hu eras', *China-US Focus*, 6 September 2022.

10   The 'New Silk Roads' are designed precisely for the purpose of securing China's supply of strategic raw materials for its own development. The

western province of Xinjiang and the city of Xi'an (Shaanxi province, in central China) play a pivotal role as overland transit points to China from Europe and the Middle East. In the dry port of Xi'an, the annual traffic of goods increased from 100 million tonnes per year in 2000 to 300 million in 2019. See Eric Armando, 'Understanding the Energy Silk Roads', Groupe d'études géopolitiques, September 2021, and X. Chen, and Z. Li, *Xi'an: From an ancient world city to a 21st-century global logistics centre*, Routledge Handbook of Asian cities, Routledge (London), 2023, pp. 172-185.

11   Information report no. 349 (2010-2011) on the security of France's strategic supplies, prepared for the French Committee on Foreign Affairs, Defence and Armed Forces by Jacques Blanc, French Senate, 2011.

12   Generally speaking, Western countries took a position of reliance on major mineral-producing countries. Accordingly, the United States is 100 per cent dependent for fifteen minerals including arsenic, indium, scandium, and manganese. It is more than 80 per cent dependent for twenty-eight minerals, and 50 per cent dependent for fifty-one minerals. See '2022 US net import reliance' (figure 2), Mineral Commodity Summaries 2023, US Department of the Interior, US Geological Survey, p. 7.

13   See *Statistical Review of World Energy, 2022*, 71st edition, BP, 2022.

14   Interview with Dudley Kingsnorth, professor at Curtin University in Australia, 2016. While China mines 'only' 68 per cent of rare earths, it controls all refining of rare earths including europium, gadolinium, terbium, dysprosium, erbium, yttrium, holmium, thulium, ytterbium, and lutetium, and up to 85 per cent of rare earths including lanthanum, cerium, praseodymium, neodymium, and samarium. See 'Study on the critical raw materials for the EU 2023', op. cit., Appendix 5.

15   Refer to the figures reported by John Seaman in 'Rare Earths and Clean Energy: analyzing China's upper hand', *Institut français des relations internationales* (IFRI), September 2010.

16   J. Korinek and J. Kim, 'Export Restrictions on Strategic Raw Materials and their Impact on Trade', OECD Trade Policy Papers, no. 95, OECD Publishing, 2010.

17   Refer to complaints DS431, DS432, and DS433, DS508 and DS509, brought to the WTO between 2012 and 2016 by the United States, the European Union, and Japan with respect to China's application of export restrictions on rare earths and numerous other metals.

18   Interview with Jean-Yves Dumousseau, 2016.

19   See also Robert L. Paarlberg, 'Lessons of the Grain Embargo', *Foreign Affairs*, Fall 1980.

20   'Russia stokes tensions with the west by cutting gas exports to Poland', *The Guardian*, 10 September 2014.

21   'As European Leaders Visit Kyiv, Putin Cuts Their Gas Supply', *The New York Times*, 16 June 2022.

22   The video is available on YouTube at https://youtu.be/sVVM2AmvD5U.

23   Interview with Toru Okabe, professor at the University of Tokyo, 2011.

24 'Amid Tension, China Blocks Vital Exports to Japan', *The New York Times*, 22 September 2010.
25 'Continental AG, Bosch Push EU to Secure Access to Rare Earths', *Bloomberg*, 1 November 2010.
26 Hillary Clinton added that 'our countries, and others will have to look for additional sources of supply. This served as a wake-up call.' 'Clinton Hopes Rare Earth Trade to Continue Unabated', *Reuters*, 28 October 2010.
27 The French Strategic Metals Committee (COMES) is a forum for consultation between ministries, public bodies, manufacturers, and trade federations. It is chaired by the minister responsible for raw materials (in 2023 this was Bruno Le Maire, France's finance minister). In 2022 an interministerial delegation for the supply of strategic minerals and metals was formed and placed under the remit of the Prime Minister. This delegation is linked to COMES, for which it now acts as secretariat. See Decree no. 2011–100 of 24 January 2011 creating the Strategic Metals Committee (COMES), as well as Decree no. 2022–1550 of 10 December 2022 on the Interministerial Delegation for strategic minerals and metals supply, available on the Légifrance website (French only).
28 Interview with John Seaman, researcher at the French Institute of International Relations (IFRI), 2015.
29 A kilogram of terbium was soon trading at €2,900 — that is, ten times higher than two years before. Source: 'Rhodia renouvelle ses filons de terres rares' ['Rhodia Renews Its Rare Earths Arm'], *L'Expansion*, 2 November 2011. In mid-2011, a kilogram of dysprosium traded at an astronomic price of nearly $3,000 — that is, 100 times more than in 2003. Source: 'Les matières premières comme enjeu stratégique majeur: le cas des "terres rares"' ['Commodities as a Major Strategic Challenge: the case of rare earths'], presentation by Christian Hocquard, commodity analyst and economic geologist at the French Geological Survey (BRGM), at the 23rd Globalisation Summit on raw materials, rare metals, and energy resources, Centre d'analyse stratégique, Paris, 5 October 2011.
30 'Had a $2 cup of coffee undergone the same inflation as europium, today it would be worth $24.55,' explains the company General Electric to its customers to explain the increases in its tariffs. Source: 'Rhodia renouvelle ses filons de terres rares', op. cit.
31 According to an OECD analysis, in 2023 China applied export taxes to aluminium, antimony, copper, nickel, phosphates, tin, tungsten, and zinc. It imposed export restrictions notably on rare earths and on metals of which it is the world's leading producer, yet a small exporter. For example, it produces 66 per cent of the world's vanadium, but accounts for just 11 per cent of global exports. By contrast, European countries, the United States and Australia do not place any restrictions on the export of raw materials. See 'Trade in raw materials, 2022 edition' and 'Compare your country' for the OECD's map of restrictions by country.
32 In 2016, Royal Bafokeng Holdings decided to reduce their share to 6.3 per cent to diversify their source of revenue, before selling their shares in Impala

Platinum in 2019. See 'RBH Portfolio Investment Timeline' under 'Who we are' on the Royal Bafokeng Holdings (RBH) website, bafokengholdings.com. Consulted in June 2023.

33 Semane Molotlegi died in 2020.

34 Thanks to their revenue, the Bafokeng have embarked on a large-scale investment campaign in a variety of sectors, including energy, insurance, telecommunication, pharmaceuticals, sports, and public works and construction. According to Royal Bafokeng Holdings, the value of their portfolio grew from 22 billion rands (around €1 billion) in 2006 to 46 billion rands (around €2.3 billion) in 2021. To 'crown' it all, they have designed a plan called Vision 2035 — the year in which platinum mining will have peaked. Thereafter, the foundations of an independent and sustainable model for the platinum economy will need to be laid. See 'What we do' on the RBH website mentioned above.

35 At a regional level, chiefs from other communities in South Africa and Zambia, and even Joseph Kabila, president of the Democratic Republic of the Congo until 2019, came.

36 From 2012, the Mongolian government promulgated a law that drastically restricts foreign investment in sectors — such as mining — that are seen as strategic.

37 See 'Resource nationalism sweeps Latin America', National Bank of Canada, 7 July 2022.

38 In 2023, BHP Billiton significantly increased its investments in Canada, and is stepping up the commissioning of a potash mine in Jansen, Saskatchewan. After merging with Agrium in 2017, PotashCorp is considering a partnership with BHP Billiton. See 'BHP to speed up $5.7 billion Jansen potash mine', mining.com, 19 July 2022, and 'BHP open to potash partnership with Nutrien, leaves door open to new takeover attempt', *The Globe and Mail*, Toronto, 14 March 2023.

39 The government eventually abandoned its position in March 2023, as the move ruffled private investors at a time when several mining companies were looking to raise funds for future operations. See 'Exclusive: Canada will not force Chinese state investors to divest stakes in Teck, First Quantum', *Reuters*, 8 March 2023.

40 By 2030, the mining sector could form the third pillar of the Saudi economy. See 'Ma'aden and Ivanhoe Electric to embark on milestone exploration program in Saudi Arabia', *International Mining*, 16 May 2023. Saudi Arabia also launched a $3 billion investment fund in 2023 to invest internationally in the mining sector, with a particular focus on critical metals such as copper, nickel, and lithium. Sources close to the fund say that the Kingdom is looking to invest up to $15 billion over the coming years. See 'Saudi Arabia launches mining fund in effort to reduce oil dependency', *The Financial Times*, 11 January 2023, and 'Saudi's PIF emerges as lead bidder for $2.5 billion Vale base metals stake', mining.com, 20 June 2023.

41 'Qatar Fund Raises Stake in Xstrata', *The New York Times*, 23 August 2012. Xstrata merged with Glencore in 2013.

42   'Qatar Mining on the offensive in Africa', *Africa Intelligence*, 1 April 2014. On
      Kazakhstan, see 'Kazakhstan and Qatar sign investment agreements worth
      $625 million', *The Astana Times*, 14 October 2022.

43   'Rising Resource Nationalism Seen as "Fire Burning" for Miners', *Bloomberg*,
      16 May 2018, and 'Resource Nationalism on the Rise in Sub-Saharan Africa',
      Mining Weekly, 15 June 2018.

44   Interview with Sukhyar Raden, General Director — Mineral & Coal for the
      Indonesian Ministry of Energy & Mineral Resources, 2015.

45   See 'Export restrictions in raw materials trade: Facts, fallacies and better
      practices', OCDE, 1 May 2014, and chart 4.1 on p. 37 of the report 'Raw
      materials critical for the green transition. Production, international trade and
      export restrictions', *OECD Trade Policy Papers*, OCDE, 11 April 2023.

46   Refer to restrictions by country as listed by the OECD in 'Trade in raw
      materials, 2022 edition', OCDE, op. cit.

47   These claims were in fact supported by Resolution 1803 (XVII) of the United
      Nations, voted in 1962, which established 'permanent sovereignty of peoples
      and nations over their natural wealth and resources'.

48   Interview with Jack Lifton, Technology Metals Research, 2016.

49   The protectionism is also essentially the outcome of Western trade practices.
      China, too, is the target of trade retaliation measures orchestrated by
      Western states and corporations. According to Global Trade Alert (GTA),
      the number of retaliatory measures taken against China more than doubled
      between 2009 and 2023. In 2022, the GTA counted no fewer than 1,500 new
      state measures taken against Beijing: 617 by the United States, and 221 by
      the European Union. See the 'Countries' section on globaltradealert.org.
      This has raised concerns from the World Economic Forum. In view of the
      rise in trade violations involving mineral resources, the Swiss foundation
      foresees four scenarios for 2030: the first is scarcity in global mineral
      resources; the second, the weakening of the economies of developing
      countries as a result of this scarcity; the third, autarky policies pursued by
      mining states that lead to resource shortages in metal-importing countries;
      and the fourth, military conflicts between states over access to the most
      strategic resources. See 'The Global Risks Report 2023', World Economic
      Forum, January 2023.

50   'Xi's visit boosts China's critical rare-earth sector', *Global Times*, 5 May 2019.

51   'Commentary: US risks losing rare earth supply in trade war', Xinhua Net,
      29 May 2019.

52   To explore this demonstration further, Sophie Lepault and Romain Franklin's
      documentary *The New World of Xi Jinping*, ARTE, 2021 (75 minutes) is
      particularly informative. Read also François Bougon's book *Inside the Mind of
      Xi Jinping*, C. Hurst & Co., 2018.

53   Global steel production stood at 1.9 billion tonnes in 2021, and rare-earth
      production at 290,000 tonnes. See *Mineral Commodity Summaries*, US
      Department of the Interior, US Geological Survey, January 2023.

54   To date, China has not signed the Global Reporting Initiative, the purpose

of which is to encourage the transparency of governments, particularly with regard to how they manage their resources.

55 'Matières premières: le grand retour des stratégies publiques' ['Raw materials: State strategies make a comeback'], *Paris Tech Review*, 4 May 2012.

56 For further understanding of the growing role of the financial sector on the commodities market, see the documentary *Commodity Traders* by Jean-Pierre Boris and Jean Crépu (2014). The following is an extract transcribed from the documentary: 'March 13th, 2000. The dotcom bubble bursts. The Nasdaq stock market index crashes. Financiers pull out of this market in search of new sources of profit. Two US economists, Gary Gorton and K. Geert Rouwenhorst, take a closer look and publish a report entitled "Facts and Fantasies about Commodity Futures". It underscores the high profitability of investing in commodities, which is also a way to diversify share portfolios. The major banks got the message and turned their sights on the commodities market.'

57 There have also been cases of speculation on palladium, cobalt, and molybdenum.

58 'Electric Carmakers on Battery Alert after Funds Stockpile Cobalt', *The Financial Times*, 23 February 2017. Another example is the massive stockpiling of indium at the end of 2009 via commodity investment vehicles to 'dry up this small market and massively inflate prices', say the press. See 'La Chine restreint ses exportations de matières premières stratégiques' ['China Limits Its Strategic Raw-material Exports'], *Le Monde*, 29 December 2009.

59 Refer to the analysis by Léane Verhulst, expert at the French Geological Survey (BRGM), 'Etude du cas de la crise du prix du nickel au LME' ['Case study on the LME nickel price crisis'], under 'Clefs d'analyse des marchés de métaux' ['Keys to reading the metals markets'] on the French mineral resources website Mineral Info. See also the website of the London Market Exchange for historical nickel prices: lme.com/Metals/Non-ferrous/LME-Nickel.

Chapter Five: High-tech hold-up

1 The metals must be smelted in precise amounts. The resulting alloy is then cooled, milled, compacted, pressed using a piston, sintered, and cooled again. For a more technical explanation, read Sandro Buss, 'Des aimants permanents en terres rares' ['Permanent Rare-earth Magnets'], *La Revue polytechnique*, no. 1745, 13 April 2010.

2 Turning the sparkwheel of the lighter strikes the flint — a blend of rare-earth metals called 'mischmetal' — producing a spark that lights the gas or the wick. As for camping lanterns, the incandescent light of the flame is produced not by gas but by the rare earth cerium. When heated by the flame, the cerium-impregnated gas mantle emits a bright white light that provides enhanced lighting. Refer to 'Terres rares: enjeux stratégiques pour le développement durable' ['Rare Earths: strategic challenges for sustainable development'], a talk by Patrice Christmann, deputy director of corporate strategy at the

French Geological Survey (BRGM), as part of a series of seminars organised by the French National Scientific Research Council (CNRS), 17 September 2013.

3   The rare earths covering the inner face of the screen are 'excited' by the cathode-ray tube to emit coloured light, and therefore images. Europium reproduces the colour red, and terbium the colour green.

4   These are the samarium-cobalt (Sm-Co) magnets, the chemical formulae of which are $SmCo_5$ and Sm_2Co_{17}, and magnets made using the rare earth neodymium, iron and boron (NdFeB), the chemical formula of which is $Nd_2Fe_{14}B$. They were invented by Masato Sagawa of Sumitomo Special Metals in Japan, and John Croat from General Motors in the US.

5   For example, samarium 'is used mainly to make permanent magnets. A technology used for the new generation of Alstom's TGVs [high-speed trains], and enabling the production of engines 30 to 40 per cent more compact, and 10 to 20 per cent more powerful.' See 'Le CAC 40 accro aux "terres rares"' ['The CAC 40's Addiction to "Rare Earths"'], *L'Expansion*, 12 November 2012.

6   Interview with Jack Lifton, 2016. According to Philippe Degobert, electrical engineering lecturer at the École nationale supérieure des Arts et des Métiers (ENSAM), and director of Masters in Mobility and Electric Vehicles (MVE), the ferrite magnet is seven times weaker than a samarium magnet, and ten times weaker than a neodymium magnet.

7   Interview with Chen Zhanheng, 2016.

8   Interview with Chris Ecclestone, 2016.

9   See Régis Poisson, 'La guerre des terres rares' ['The Rare Earths War'], *L'Actualité chimique*, no. 369, December 2012.

10   Interviews with Jean-Paul Tognet, 2016 and 2017.

11   Ibid.

12   Interview with Jean-Yves Dumousseau, 2016.

13   Régis Poisson, 'La guerre des terres rares', op. cit.

14   Interview with Jean-Yves Dumousseau, 2016.

15   Interviews with Jean-Paul Tognet, 2016 and 2017.

16   Ibid. In 2023, the La Rochelle site had 320 staff and had an annual output of less than 10,000 tonnes of rare-earth products. These include catalysts for internal combustion engine vehicles, and fine powders for surface polishing. More information on the La Rochelle site solvay.fr in the section 'Implantations' (French only).

17   Ibid. In 2022 Solvay was a multinational operating in several companies with 22,000 employees, 53 per cent of which in Europe. In China, the group has fifteen subsidiaries and nine active industrial sites out of one hundred worldwide.

18   Interview with Jim Robinson, United Steelworkers (USW), 2011.

19   Interview with Régis Poisson, former engineer at Rhône-Poulenc, 2013.

20   See Régis Poisson, 'La guerre des terres rares', op. cit.

21   Interview with David Merriman, Roskill Consultancy, 2016.

22 According to a French industrialist who preferred to speak anonymously, the purchase price ratio for European producers and Chinese producers can be as high as 1:7 — a price differential that Jean-Paul Tognet finds excessive.

23 Namely, the 'Obtaining National and Secure Homeland Operations for Rare Earth Manufacturing Act of 2023' or the 'ONSHORE Manufacturing Act of 2023'. Bill introduced in Senate in April 2022.

24 'Rubio speaks on China at American Affairs Finance conference', press release, Marco Rubio, US Senator for Florida, 7 December 2022.

25 *Rare Earth Permanent Magnets. Supply Chain Deep Dive Assessment*, US Department of Energy response to executive order 14017, 'America's Supply Chains', US Department of Energy, 24 February 2022.

26 See 'Developing a supply chain for recycled rare earth permanent magnets in the EU', CEPS (Centre for European Policy Studies, Brussels), 16 December 2022. See also 'Le marché des terres rares en 2022: filières d'approvisionnement en aimants permanents' ['The rare earth market in 2022: supply chains for permanent magnets'], *Minéral Info*, 4 July 2022.

27 Interview with Jean-Yves Dumousseau, 2011.

28 See 'Baotou rare earth high-tech zone: Optimize the characteristic industry and create a new highland of intelligent manufacturing', *Shanghai Metals Market*, 31 August 2021.

29 'Milestones of the Baotou rare earth high-tech industrial development zone in H1', *Invest in China*, 22 July 2022.

30 Interview with Dudley Kingsnorth, 2016.

31 Industrial robots are in fact the cornerstone of tomorrow's smart and ultra-connected '4.0 factories', where Germany has the lead. In 2015, the German machine tool industry generated over €15 billion, exported three-quarters of its production and had a headcount of nearly 70,000 employees. It is a pillar of the national economy. See 'German Machine Tool Industry Expects Moderate Growth in 2016', *Verein deutscher Werkzeugmaschinenfabriken*, 2016.

32 Interview with Chris Ecclestone, 2016.

33 For a more detailed explanation, see the French Geological Survey's public report: 'Panorama du marché du tungstène' ['View of the Tungsten Market'], BRGM, July 2012.

34 Interview with Chris Ecclestone, 2016.

35 The Mittelstand may have won the battle, but they did not win the war. China has a keen interest in some of Germany's star industrial robots, such as KUKA. See 'Allemagne: le "Mittelstand" face à l'offensive chinoise' ['Germany: the Mittelstand faces the Chinese offensive'], *Le Monde*, 4 June 2016.

36 Graphene's applications are astounding: bendy mobile phones, see-through laptops, ultra-powerful nanoprocessors, or nanochips that can be inserted into the human body to detect cancers, and so on.

37 'US Brings WTO Challenge against China over Copper, Graphite, Other Minerals', *The Wall Street Journal*, 13 July 2016.

38 'United States Expands Its Challenge to China's Export Restraints on Key Raw Materials', Office of the United States Trade Representative, July 2016.

39 Interview with Daisy Chen, journalist at the Beijing Bureau of Metal Pages, 2016. The case is ongoing and can be consulted on the WTO website (DS558: China — Additional Duties on Certain Products from the United States).

40 'China leads rise in export restrictions on critical minerals, OECD says', *The Financial Times*, 11 April 2023.

41 Interview with Thomas Kruemmer, 2016.

42 And why not do an inventory of 'critical super magnets' that could potentially be in short supply? It's a legitimate question since China, as part of its strategy to attract industries that use rare earths, threatened Denmark, a major producer of wind turbines, to suspend its rare-earth magnet exports, according to Chris Ecclestone.

43 Interview with Didier Julienne, 2016.

44 Interview with Jack Lifton, 2016.

45 Interview with Alain Liger, 2016.

46 'Tin: the secret to improving lithium-ion battery life', *Forbes*, 23 May 2012.

47 See *Mineral Commodity Summaries 2023*, United States Geological Survey, 2023.

48 Interview with Agung Nugroho Soeratno, head of communication at PT Timah, 2014.

49 Interview with Peter Kettle, analyst at the International Tin Research Institute, in the UK, 2016.

50 'Shanghai to Match London Metals as China Seeks Commodities Sway', *Bloomberg News*, 26 March 2015. Futures are one of the financial instruments traded on the derivatives market. To offset price instability, buyers and sellers agree on the future sale of goods based on a price set in advance.

51 'China ShFE Plans Commodities Platform to Set Physical Prices', *Reuters*, 15 May 2018.

52 'Bursa Malaysia Derivatives Introduces Futures Tin Contract', *The Star*, 29 September 2016.

53 There is no guarantee that such a policy had any bearing on global prices in the short term. Peter Kettle nevertheless maintains that 'there are clear opportunities for local brokers who benefit from financial transactions executed locally that would otherwise have taken place in the UK'. As such, stock exchanges like these actually strengthen the status of Asian cities as financial hubs.

54 Nickel and bauxite. See 'Indonesia Eases Ban on Mineral Exports', *The Financial Times*, 13 January 2017.

55 Indonesia produced 1 million tonnes of nickel 2021, or a third of the world's estimated output of 2.7 tonnes. Refer to the section on nickel in the *Mineral Commodity Summaries 2023* of the United States Geological Survey. See also 'Investments in Indonesia's nickel industry', *Reuters*, 6 February 2023.

56 Hervé Kempf, *Fin de l'Occident, naissance du monde* [*End of the West, Birth of the World*], Seuil, 2013.

57 From a speech given at a conference organised by Cyclope Circle in Paris 2016.

58 See the African Union's February 2009 'Africa Mining Vision' report, and

'Second Continental Report on the implementation of Agenda 2063', African Union Development Agency, February 2022.

59 See 'Sub-Saharan Africa minerals exports by region in US$ thousand 2020', World Integrated Trade Solution (WITS) on the World Bank website (wits. worldbank.org).

60 Refer to the World Bank Report, 'Africa's Resource Future: Harnessing Natural Resources for Economic Transformation during the Low-Carbon Transition', International Bank for Reconstruction and Development / The World Bank, 2023.

Chapter Six: The day China overtook the West

1 Claude Chancel and Libin Liu Le Grix, *Le Grand Livre de la Chine* [*The Big Book of China*], Eyrolles, 2013.

2 James McGregor, APCO senior counsellor, describes China's National Medium- and Long-Term Plan for the Development of Science and Technology (MLP) as the 'grand blueprint of science and technology development' to bring about the 'great renaissance of the Chinese nation'. James McGregor, China's Drive for 'Indigenous Innovation' — A Web of Industrial Policies, US Chamber of Commerce, 27 July 2010.

3 The industries prioritised were energy conservation and environmental protection, next-generation IT, biotechnology, high-end equipment manufacturing, new energy, new materials, and clean-energy vehicles. See 'China's 12th Five-Year Plan: how it actually works and what's in store for the next five years', APCO Worldwide, 10 December 2010.

4 Refer to the summary of the 13th Five-Year Plan in 'Prosperity for the masses by 2020 – China's 13th Five-Year Plan and its business implications', PwC China, Hong Kong and Macau, 2015.

5 See 'Outline of the People's Republic of China 14th Five-Year plan for national economic and social development and long-range objectives for 2035', Center for Security and Emerging Technology (Georgetown University, Washington), 13 May 2021.

6 Interview with Xue Lan, 2016.

7 I have borrowed this breakdown from Malo Carton and Samy Jazaerli's book, *Et la Chine s'est éveillée. La montée en gamme de l'industrie chinoise* [*And Then China Stirred: Chinese industry moves upmarket*], Presses des Mines, 2015.

8 Interview with Ding Yifan, researcher at the Institute of World Development, 2016.

9 The national medium- and long-term plan for the development of science and technology (2006–2020), The State Council of the People's Republic of China, 2006.

10 James McGregor, 'China's Drive for "Indigenous Innovation" — A Web of Industrial Policies', op. cit.

11 Ibid.

12 'Protecting American Intellectual Property Act of 2022', introduced by Senator Chris Van Hollen, and signed by President Biden on 5 January 2023.

13   See Jean-Louis Beffa, *Les Clés de la puissance* [*The Keys to Power*], Seuil, 2015.

14   The full name is '1986 National High Technology Research and Development Program'. The first two numbers of Program 863 refer to the year the program was started, and the third number to the month (March).

15   These industries are information technology, biology, aerospace, automation, energy, materials, and oceanography.

16   '2022 Global Funding Forecast: R&D variants cover more than the pandemic', *Research & Development World*, 12 April 2022, and 'EU investment in R&D increased to €328 billion in 2021', *Eurostat*, 29 November 2022.

17   Interview with Bo Chen, professor at the Shanghai Free Trade Zone research institute, 2016.

18   'China "Employs 2 Million to Police Internet"', *CNN*, 7 October 2013. See also Ryan Fedasiuk's *Buying silence: The price of internet censorship in China*, The Jamestown Foundation (Washington), 12 January 2021.

19   See Patrick Allard, 'La Chine, championne technologique ou géant empêtré' ['China, technology champion or hobbled giant'], *Politique étrangère no.1/2020*, Armand Colin, Spring 2021.

20   Interview with Bruno Gensburger, 2016. One French expert I spoke to who preferred not to be quoted said he heard the following remark from the lips of the Chinese director of an electronics group in reference to his staff: 'They don't have ideas because they are obedient. And if they're not obedient, I hit them.'

21   Bernard Apremont, 'L'économie de l'URSS dans ses rapports avec la Chine et les démocraties Populaires' ['The Economy of the USSR in Its Relations with China and Popular Democracies'], *Politique étrangère*, 1956, vol. 21, no. 5, pp. 601–13.

22   James McGregor, 'China's Drive for "Indigenous Innovation" – A Web of Industrial Policies', op. cit.

23   See 'Urban and rural population of China from 2012 to 2022', *Statista*, January 2023, and 'China creates 12,06 mln new urban jobs in 2022 – state media', *Reuters*, 10 January 2023.

24   Interview with Julien Girault, journalist at the Agence France-Press Beijing desk, 2016.

25   Interview with Ding Yifan, 2016.

26   Simone McCarthy, 'China sends first civilian astronaut to space as Shenzhou-16 blasts off', *Cable News Network (CNN)*, 31 May 2023.

27   'China plans to land astronauts on moon before 2030', *Politico*, 29 May 2023.

28   China is still second place to the US, which launched eighty-seven orbital missions in 2022. See 'Les lancements orbitaux américains sur l'année 2022 et capacités associées' ['US orbital launches in 2022 and related capacities'], *France-Science*, French Embassy in the United States, 10 May 2023.

29   Interview with Xue Lan, 2016. This progress heralded the growth in developing countries' contribution to the production of skills vis-à-vis the duopoly held by the US and Europe. A new 'duel of intelligence' (to use the expression coined by Claude Chancel and Libin Liu Le Grix in *Le Grand Livre de la Chine*, op. cit.) that Irina Bokova, the then director-general of UNESCO, highlighted

in the 2015 UNESCO report on science. In it, she stated that 'the North–South divide in research and innovation is narrowing, as a large number of countries are incorporating science, technology and innovation in their national development agendas.' See 'UNESCO Science Report: Towards 2030', 2015. Five years earlier, Bokova also stated: 'The bipolar world in which science and technology (S&T) were dominated by the Triad made up of the European Union, Japan and the USA is gradually giving way to a multi-polar world, with an increasing number of public and private research hubs spreading across North and South.' See 'Research and Development: USA, Europe and Japan increasingly challenged by emerging countries, says a UNESCO report', *Unescopress*, 10 November 2010.

30 See 'Key IP5 statistical indicators 2022', Five IP Offices.

31 'J-20: The stealth fighter that changed PLA watching forever', *The Diplomat*, 11 January 2021.

32 According to the Top500 table of supercomputer statistics. In June 2023, the American supercomputer 'Frontier' dominated the ranking. China's 'Sunway TaihuLight' is in seventh position. See 'June 2023', Lists, *Top500.org*, June 2023.

33 'China launches world's first quantum communications satellite', *New Scientist*, 16 August 2016.

34 'ASPI's critical technology tracker', Australian Strategic Policy Institute (ASPI), 1 March 2023; 'China Trumps US in Key Technology Research, Report Says', *The Wall Street Journal*, 2 March 2023.

35 *Solar PV Global Supply Chains*, International Energy Agency (IEA), July 2022.

36 'Largest hydroelectric power generating countries worldwide in 2021', *Statista*, 2023.

37 *Renewable Capacity Statistics 2023*, International Renewable Energy Agency (IRENA), March 2023.

38 *Global EV outlook 2023*, IEA, April 2023.

39 *E-Mobility Index 2021*, Roland Berger (think tank on advanced technology), March 2021.

40 'Top highlights of world EV sales in 2022', *Clean Technica*, 15 March 2023. BYD launched an offensive on the global markets; Europe, Australia, South Amerca, and South-East Asia (see 'BYD powers on', *Seeking Alpha*, 12 July 2023), and has now overtaken Tesla as the world's biggest EV maker: 'A Chinese EV Company Has Taken Tesla's Crown', *Foreign Policy*, 4 January 2024.

41 'US announces steps to bolster critical mineral supply chain', *Voice of America*, 22 February 2022.

42 'Can the world make an electric car battery without China?', *The New York Times*, 16 May 2023.

43 See 'Red China's Green Crisis', *Le Monde diplomatique*, July 2017.

44 These are the key takeaways of the research by Karl Gerald Van den Boogaart from the Freiberg University of Mining and Technology. His work was presented at the Denver SME Critical Minerals Conference in 2014 and quoted by Dudley Kingsnorth at the 5th Annual Cleantech and Technology Metals Summit in Toronto in April 2016.

45 Dudley Kingsnorth, professor at Curtin University, Western Australia.

46 'Global rare earth metals market to register a staggering 7.4% CAGR from 2022 to 2031, reaching US$ 21.7 billion: TMR Report', Transparency Market Research, *Globe NewsWire*, 6 June 2023. The global crude oil market was estimated at over $2 trillion in 2022. 'Sizing up: The oil market vs top 10 metal markets combined', *Visual Capitalist (VC) Elements*, 30 June 2023.

47 Interview with Peter Dent of the Electron Energy Corporation, 2011.

48 Interview with Dudley Kingsnorth, 2016.

49 'Giga Shanghai leads Tesla production, capacity surpasses 750K EVs per year', *Pandaily* (Beijing), 21 July 2022.

50 Interview with Dudley Kingsnorth, 2016.

51 'Donald Trump Hails New Era of US Energy "dominance"', *The Financial Times*, 30 June 2017.

52 These political choices are nevertheless based on the false assumption according to which the US alone can direct new energy balances in one direction rather than another, whereas the Chinese now hold the cards. In addition, by refusing to tackle Beijing head-on, Washington has already admitted defeat.

53 Arnaud Montebourg, 'L'Europe ne peut plus être à ce point désinvolte sur la mondialisation' ['Europe Cannot Continue to Be so Lackadaisical about Globalisation'], *Le Monde*, 26 October 2016.

54 Interview with Didier Julienne, 2015.

55 Interview with Gary Hubard, United Steelworkers, 2011.

56 The share decreased from 16.4 per cent to 12.4 per cent. Refer to the 'Industry in France' infographic on www.gouvernement.fr, 2 April 2015. Note, however, the strong figures from the French industrial sector in 2017. See 'La France recrée enfin des usines' ['France Is Finally Recreating Factories'], *Le Monde*, 29 September 2017.

57 'Industry (Including Construction), Value Added (Per Cent of GDP)', The World Bank, 2019.

58 Jean-François Dufour, *Made by China: les secrets d'une conquête industrielle* [*Made in China: the secrets of an industrial conquest*], Dunod, 2012.

59 Brahma Chellaney, 'The Challenge from Authoritarian Capitalism to Liberal Democracy', *China-US Focus*, 6 October 2016.

60 The expression was suggested by Joshua Cooper Ramo in a 2004 academic paper of the Foreign Policy Centre entitled 'The Beijing Consensus'.

61 Zhao Tingyang, *The Tianxia System: an introduction to the philosophy of a world institution*, Jiangsu Jiaoyu Chubanshe, 2005.

62 Interview with Chen Zhanheng, 2016.

Chapter Seven: The race for precision-guided missiles

1 In nuclear power plants, samarium-149 is used to absorb neutrons and thus reduce the reactivity (the rate of fission) of nuclear fuel. Contrary to the fiction, it is not a rare earth in itself, but an isotope of samarium.

2 Numerous reports investigate the importance of rare earths in the US defence

industries. See the chapter 'Review of critical minerals and materials — Department of Defense' in the document 'Building Resilient Supply Chains, Revitalizing American Manufacturing, and Fostering Broad-Based Growth', The White House, June 2021. A more recent report by a Dutch research institute studies the needs of the European defence industry: 'Strategic Raw Materials for Defence Mapping European Industry Needs', The Hague Centre for Strategic Studies, January 2023.

3 This type of hybrid conflict combines conventional military means with misinformation strategies aimed at influencing public opinion. See 'The other front of the Russian offensive is a hybrid war aimed at weakening the EU and democracy', *Le Monde*, 15 March 2023.

4 Jean-Claude Guillebaud, *Le Commencement d'un monde. Vers une modernité Métisse* [*The Beginning of a World: towards hybrid modernity*], Seuil, 2008.

5 Interview with Jack Lifton, Technology Metals Research, 2016.

6 'Department of Defense Releases the President's Fiscal Year 2024 Defense Budget', press release of the US Department of Defense, 13 March 2023.

7 'Perpetua Resources gets DOD funding to study antimony production from Stibnite gold project', mining.com, 20 September 2022.

8 For the full story on the delocalisation of Magnequench, read Charles W. Freeman III, 'Remember the Magnequench: an object lesson in globalization', *The Washington Quarterly*, The Center for Strategic and International Studies, 2009, pp. 61–76.

9 Consult the database of the Stockholm International Peace Research Institute (Sipri): SIPRI Military Expenditure Database, milex.sipri.org/sipri.

10 See Thierry Sanjuan, 'L'Armée populaire de libération: miroir des trajectoires modernes de la Chine' ['The People's Liberation Army: mirror on the modern trajectories of China'], *Hérodote*, no. 116, 2005.

11 Mycenaean Greece owed its prosperity to its military superiority over its enemies — particularly the Trojans — which it acquired from such weapons.

12 The Incas' and other Andean cultures' mastery of copper and bronze was no match for the Conquistadors' mastery of iron. This contributed to the swift conquest of the Americas in the fifteenth and sixteenth centuries. See Eric Chaline, *50 Minerals that Changed the Course of History*, Firefly Books, 2012.

13 Nabeel Mancheri, Lalitha Sundaresan, and S. Chandrashekar, 'Dominating the World: China and the rare earth industry', *National Institute of Advanced Studies*, Bangalore, India, April 2013.

14 'Panama Papers: from the Guatemalan Drug Queen to the "most dangerous mobster in the world"', *The Irish Times*, 9 May 2016.

15 The Watergate scandal led to the resignation of President Richard Nixon in 1974.

16 'China's First Family Comes Under Growing Scrutiny', *The New York Times*, 2 June 1995.

17 The 'four commandments' governing the policy are: 'Combine the military and civil', 'Combine peace and war', 'Give priority to military products' and 'Let the civil support the military'. Together, they form sixteen ideograms, hence the name 'sixteen characters'.

18   'China's Spies "Very Aggressive" Threat to US', *The Washington Times*, 6 March 2007.

19   Interview with Hugo Meijer, then a researcher at L'Institut de recherche stratégique de l'École militaire (IRSEM), 2016.

20   Interview with Steve Constantinides, Arnold Magnetic Technologies, 2016.

21   Ibid.

22   Scott Wheeler, 'Trading with the Enemy: how Clinton administration armed communist China', *American Investigator (Free Republic)*, 13 January 2000.

23   'Illegal Fundraiser for the Clintons Made Secret Tape because He Feared Being ASSASSINATED over What He Knew – and Used It to Reveal Democrats' Bid to Silence Him', *The Daily Mail*, 23 February 2017.

24   'Democrats Return Illegal Contribution; Politics: South Korean subsidiary's $250,000 donation violated ban on money from foreign nationals', *The Los Angeles Times*, 21 September 1996.

25   'Chinese Embassy Role in Contributions Probed', *The Washington Post*, 13 February 1997.

26   The conclusions of the enquiry were released to the public in April 2019. 'G.O.P.-led senate panel details ties between 2016 Trump campaign and Russia', *The New York Times*, 18 April 2020.

27   The Pacific fleet comprises 200 ships and 150,000 US Marine Corps personnel. With 350 ships positioned in the same area, China has increased the size of its fleet tenfold in twenty years. See 'The state of the US Navy as China builds up its naval force and threatens Taiwan', *CBS News*, 2 July 2023.

28   See 'La mer de Chine méridionale, l'autre versant de la rivalité sino-américaine en Asie' ['The South China Sea: a Major Frontier Issue in Southeast Asia'], *Le Rubicon*, 13 October 2022. The European Parliament provides an excellent brief overview of the maritime issues and strategies of China, India, and the United States in the Indo-Pacific region. See 'Geopolitics in the Indo-Pacific: Major players' strategic perspectives', Briefing for the European Parliament – European Parliamentary Research Service, July 2023.

29   Global military spending in 2022 totalled $2,200 billion. See *SIPRI Military Expenditure Database*, op. cit.

30   'Biden says US forces would help defend Taiwan', *Le Monde*, 19 September 2022.

31   'Presidential Executive Order 13806 – Assessing and Strengthening the Manufacturing and Defense Industrial Base and Supply Chain Resiliency of the United States', The White House, 21 July 2017.

32   'It's Not Buy America: admin aide on Trump's sweeping industrial base study', *Breaking Defense*, 25 July 2017.

33   'US Launches National Security Probe into Aluminum Imports', *Reuters*, 27 April 2017.

34   'Building Resilient Supply Chains, Revitalizing American Manufacturing, and Fostering Broad-Based Growth', The White House, June 2021.

35   'Pentagon bankrolls rare earths plant as US plays catch-up to China', *The Financial Times*, 14 March 2022. See also 'MP Materials begins construction

on Texas rare earth magnetics factory to restore full US supply chain', press release, 21 April 2022.

36   This law was broadened in 2009 to include electromagnets.

37   'F-35's new $412 billion cost estimate is a modest increase by its standards', *Bloomberg*, 28 September 2022.

38   'F-35 production challenged to keep up with demand', *Air and Space Forces Magazine*, 3 July 2023.

39   'Defense Science Board Task Force on High Performance Microchip Supply', Office of the Under Secretary of Defense for Acquisition, Technology, and Logistics, Defense Science Board (DSB), United States, 2005.

40   'Exclusive: US waived laws to keep F-35 on track with China-made parts', *Reuters*, 3 January 2014.

41   'F-35s all contain banned China-made alloy, Pentagon says', *Bloomberg*, 9 September 2022.

42   'With Pentagon waiver, deliveries resume of new F-35s with Chinese magnets', *Air and Space Forces Magazine*, 11 October 2022.

43   Public Law 117-263, 117th Congress (United States), 23 December 2022.

44   A great many companies are relocating permanent magnet production to the US: MP Materials, Quadrant Magnetics, USA Rare Earth, and Noveon — a company that innovates in magnet recycling (its director estimates that production will be at full capacity, i.e. 2,000 tonnes of magnets annually, in 2024 or 2025). German company Vacuumschmelze, a rare-earth magnet manufacturer since 1973, also has plans to open a plant in the United States in 2026. See John Ormerod, 'Reshoring of the US rare earth magnet industry', *Knowledge Ridge*, 7 April 2023, and 'Meet the Texas startup that recycles rare-earth magnets, bypassing China', *Forbes*, 10 May 2023. Visit the Vacuumschmelze website for more information on the types of rare-earth permanent magnets and their composition (vacuumschmelze.com/products/permanent-magnets).

Chapter Eight: Mining goes global

1   These essential minerals include aluminium, lead, iron, copper, nickel, chrome, and zinc. Source: Data Series 140, United States Geological Survey (USGS), 2017.

2   Jean-Baptiste Fressoz, 'The Anthropocene as an "accumulocene"', *Regards croisés sur l'économie*, vol. 26, Issue 1, January 2020.

3   'What will the global economy look like in 2050?', Capital Economics, 2023.

4   Alain Liger, Secretary General of the French Strategic Metals Committee (COMES), 'Transition énergétique: attention, métaux stratégiques!' ['The Energy Transition: it's about strategic metals!'], The French High Council for Economy, Industry, Energy and Technology, 7 December 2015.

5   'The Role of Critical Minerals in Clean Energy Transitions', International Energy Agency (IEA), May 2021.

6   Since the production of steel and copper — two major base-metal products — was stable between 1970 and 2000, there was no real concern about the

possibility of a mineral shortage. It was only in 2005 that industrial players and the press began to talk about scarcity after the sudden emergence of China on the raw materials market, which put immense strain on supplies.

7   'Ressources minérales et énergie: rapport du groupe Sol et sous-sol de l'Alliance' ['Mineral Resources and Energy: report from the Alliance's ground and underground group'], Alliance nationale de coordination de la recherche scientifique (ANCRE), June 2015. For further reading, also refer to Olivier Vidal's book *Mineral Resources and Energy: future stakes in energy transition*, ISTE Press Ltd, 2018.

8   Olivier Vidal, Bruno Goffé, and Nicholas Arndt, 'Metals for a Low-Carbon Society', *Nature Geoscience*, vol. 6, November 2013.

9   Ibid.

10   'The Growing Role of Minerals and Metals for a Low Carbon Future', The World Bank Group, June 2017. See also 'Métaux: les besoins colossaux de la transition énergétique' ['Metals: the colossal needs of the energy transition'], *Les Échos*, 20 July 2017. These conclusions were confirmed by the IEA in its report *The Role of Critical Minerals in Clean Energy Transitions* (May 2021). The report finds that an electric vehicle, over its entire lifecycle, requires around six times more metals and minerals than internal-combustion engine vehicles. Every megawatt of electricity generated by an offshore wind turbine requires ten times more resources than the same output of a gas-fired plant. The electricity generated by a solar panel requires two-and-a-half more resources than the electricity generated by a coal-fired plant.

11   'How Many People Have Ever Lived on Earth?', Population Reference Bureau, 15 November 2022.

12   This very issue of water consumption is addressed in the study 'Without water, can we surf the internet tomorrow?' in the research journal *Humanité et numérique: Les liaisons dangereuses*, coordinated by Dr Servane Mouton, Editions Apogée, 'Les panseurs sociaux' collection, 2023.

13   Interview with Olivier Vidal, 2023. Alain Liger, who organised the rare metals symposium at the COP 21 was equally disappointed: 'I sent a note to Ségolène Royal [the then French environment minister], Emmanuel Macron [the then French finance minister], and Laurent Fabius [the then French foreign affairs minister]. Macron's staff called to congratulate me on the event. But I didn't hear back from Fabius or Royal' — the very ministers leading the climate talks.

14   The title of the panel discussion was 'Critical minerals for Net Zero — How to ensure sustainability and circularity'. It was organised by Y (Youth) 20 Indonesia and the BRICS (Brazil, Russia, India, China, South Africa) Youth Energy Agency, and moderated by Vadim Kuznetsov — a 'young delegate' at the time for climate and energy at the G20 meeting in Indonesia in 2022.

15   One need only refer to the alarming words of environmentalist and European member of parliament Yannick Jadot who would rather 'depend on wind, sun and water, than on Putin and the petro-monarchies of the Gulf'. What he omits entirely is that between wind and the electrical plug is metals-based green tech. See 'Transition énergétique: EELV veut 100 milliards d'euros par

an de l'Europe' ['Energy transition: the French Greens want 100 billion euros a year from Europe'], *Sud-Ouest*, 20 August 2018.

16   Emmanuel Hache, *Metals in the energy transition*, IFP Energies nouvelles (IFPEN), 2023.

17   'Fer, cuivre, aluminium... les ressources minières seront épuisées d'ici à 50 à 70 ans si nos consommations continuent de croître au rythme actuel.' ['Iron, copper, aluminium... mining resources will be depleted in the next 50 to 70 years at the current growth rate of consumption'], *L'Usine nouvelle*, 23 May 2023.

18   Interview with John Petersen, 2017.

19   John Seaman, 'Rare earths and China: A review of changing criticality and the new economy', *French Institute of International Relations*, January 2019.

20   'Ruée mondiale vers le cobalt congolais: la Chine, médaille d'or' ['Global scramble for Congolese cobalt: China takes the gold medal'], *Capital*, 21 February 2018. See also 'Confrontation autour du cobalt congolais' ['Confrontation over Congolese cobalt'], *Ecole de guerre économique (EGE)*, 1 June 2023.

21   Paul Valéry, *Regards sur le monde actuel*, Librairie Stock, Delamain et Boutelleau, 1931.

22   Donella H. Meadows, Dennis L. Meadows, Jorgen Randers, and William W. Behrens III, *The Limits to Growth: a report for the Club of Rome's project on the predicament of mankind*, Universe Books, 1972.

23   Jean-Baptiste Say, *A Treatise on Political Economy*. Translated by C.R. Prinsep. Ontario: Batoche Books, 1880.

24   'Almost 400 new mines needed to meet future EV battery demand, data finds', *Tech Informed*, 27 September 2022.

25   WMF [World Materials Forum] criticality assessment by BRGM, CRU & McKinsey, presented by Christophe Poinssot, 6 July 2023.

26   'Greenland says no to China-backed rare-earth mine in election', *Nikkei Asia*, 8 April 2021.

27   'Serbia pulls plug on Rio Tinto's $2.4 billion lithium project', *CNN*, 21 January 2022.

28   'Europe's green dilemma: Mining key minerals without destroying nature', *Politico*, 15 March 2023.

29   Namely, the Shevchenkivske deposit, near Donetsk in the Donbas region, and the Dobra deposit in the Kirovograd region in western Ukraine. See 'Russia-Ukraine war risks hampering transition to global low-carbon economy', *Energy Monitor*, 20 July 2023.

30   'BHP Group to ramp down Chilean copper mine after court's water extraction ban', *S&P Global Market Intelligence*, 5 January 2022.

31   'Drought in China's Yunnan set to cap province's aluminium output', *Reuters*, 20 April 2023.

32   Ugo Bardi, *Extracted: how the quest for mineral wealth is plundering the planet*, Chelsea Green Publishing, 2014.

33   Siddharth Misra, 'Plummeting "energy return on investment" of oil and the impact on global energy landscape', *Journal of Petroleum Technology (JPT)*, 20 March 2023.

34 Gonzalo Chiriboga, Andrés de la Rosa, Camila Molina, Stefany Velarde and Ghem Carvajal, 'Energy return on investment (EROI) and life cycle analysis (LCA) of biofuels in Ecuador', *Heliyon*, vol. 6, no. 6, 2020.

35 So says Christian Thomas, company founder of the metals recycling company Terra Nova.

36 Ester van der Voet et al., *Environmental Risks and Challenges of Anthropogenic Metals Flows and Cycles: a report of the working group on the global metal flows to the International Resource Panel*, Kenya, United Nations Environment Programme (UNEP), 2013.

37 Interview with Mark Cutifani, then Chief Executive of Anglo American, 2021. Cutifani added: 'Anglo American aims to reduce 30 per cent of its energy consumption per unit of copper produced by 2030. This means more precise extraction of minerals, more efficient rock grinding and selective processing, new metal recovery methods in the floating processes, and using less energy in eliminating waste. We're also trying to halve our water consumption in the process, which would bring our water consumption performance per unit of copper produced to 1900 levels.'

38 Simon P. Michaux, 'The Mining of minerals and the limits to growth', Geological survey of Finland (GTK), 1 March 2021.

39 Interviews with Jean-Paul Tognet, chemical engineer in mining and metals, 2016 and 2017.

40 Bardi, *Extracted*, op. cit.

41 'Clean energy demand for critical minerals set to soar as the world pursues net zero goals', IEA press release, 5 May 2021.

42 Simon P. Michaux, 'Challenges and bottlenecks for the green transition', Geological Survey of Finland (GTK), 21 February 2023. These statements form the conclusions of Michaux's report for the GTK entitled 'Assessment of the extra capacity required of alternative energy electrical power systems to completely replace fossil fuels', published on 20 August 2021.

43 Simon Winchester, *The Map that Changed the World: William Smith and the birth of modern geology*, HarperCollins, 2001.

44 Refer to the second half of the commission of enquiry report of the French Senate, 'Santé publique: pour un nouveau départ. Leçons de l'épidémie de covid-19' ['Public health: a new departure. Lessons learned from the Covid-19 epidemic'], 8 December 2020.

45 'Suez Canal blockage adds to global supply chain chaos', *The Sydney Morning Herald*, 25 March 2021.

46 'A dozen EU countries affected by Russian gas cuts, EU climate chief says', *Reuters*, 23 June 2022.

47 Speech by M. Emmanuel Macron, President of the Republic, at the Nexus Institute (The Hague, 11/04/2023), Embassy of France in Washington, DC.

48 H.R.5376-Inflation Reduction Act of 2022, 117th Congress (2021-2022) (congress.gov/bill/117th-congress/house-bill/5376).

49 'Biden invokes cold war statute to boost critical mineral supply', *The New York Times*, 31 March 2022.

50   2022 State of the Union Address by President von der Leyen, 14 September 2022, available on the website of the European Commission (europa.eu).

51   'Philippe Varin: "We need to develop a real metals diplomacy"', French Geological Survey (BRGM), 5 July 2022. The report is entitled 'Investir dans la France de 2030: remise au gouvernement du rapport Varin sur la sécurisation de l'approvisionnement en matières premières minérales et ouverture d'un appel à projets dédié' ['Invest in the France of 2030: submission to the government of the Varin report on securing the provision of mineral raw materials, and on launching an appeal for projects to achieve this endeavour'], 10 January 2022. It is available on the website of the French government: gouvernement. fr. Philippe Varin held various positions in groups including Pechiney, Corus, PSA and Suez.

52   'Global rare earth element (REE) mines, deposits and occurrences (May 2021)', British Geological Survey.

53   Patricia Schouker, 'Trillions at stake: The next battle in Asia is over north Korea's rare earth resources', *The National Interest*, 6 February 2020.

54   'Is the US subverting China's influence in the DRC?', *ThinkChina*, 1 March 2023.

55   'Germany, Canada to boost energy, mineral ties as they decarbonize', *Reuters*, 22 August 2022.

56   'Macron Says Mongolia to Supply Critical Metals for Green Push', *Bloomberg*, 21 May 2023.

57   'US, Australia, Japan, India to collaborate on rare earths production', *S&P Global Market Intelligence*, 11 March 2021.

58   Minerals Security Partnership. Available on the website of the US Department of State (14 June 2022): state.gov/minerals-security-partnership.

59   'I have just signed a bill to reform the mining law to make lithium a mineral that belongs to the nation [...]. Lithium, much sought after by multinationals and foreign governments, will no longer be up for grabs', stated Mexico's president before the vote in parliament. See 'Mexican president reignites debate around mining reform with focus on lithium', mining.com, 10 April 2022.

60   See 'Chile passes mining royalties reform after five years of debate', *The Brazilian Report*, 19 May 2023, and 'Brazil preparing for a new mining plan', *The Assay*, February 2023.

61   'Emerging lithium producer Ghana will avoid exporting raw mineral', *Bloomberg*, 14 June 2023.

62   'The hidden gigafactories powering emerging markets', *Benchmark Source*, 13 December 2022.

63   'The transition to clean energy will mint new commodity superpowers', *The Economist*, 26 March 2022.

64   Interview with Didier Julienne, natural resources strategist, 2016.

65   'Why energy insecurity is here to stay', *The Economist*, 26 March 2022.

66   'Secretary Blinken at an MOU Signing with Democratic Republic of the Congo Vice Prime Minister and Foreign Minister Christophe Lutundula and

Zambian Foreign Minister Stanley Kakubo', press release, US Department of State, 13 December 2022.

67   Refer to the fascinating paper by Mark Leonard et al. for the Bruegel Institute, 'The geopolitics of the European Green Deal', *Policy Contribution*, Issue no. 04/21, February 2021. To the best of my knowledge, one of the very first reports published on the geopolitics of the energy transition is 'The Geopolitics of Renewable Energy' by Meghan O'Sullivan, Indra Overland and David Sandalow (HKS Working Paper no. RWP17-027), Harvard Kennedy School Faculty Research Working Paper Series / Columbia University Center on Global Energy Policy, July 2017.

68   'Our work in Africa', United Nations Environment Program, unep.org.

69   Read 'Chinese foreign mining investment. China's private sector eyes low-cost regions', *S&P Global Market Intelligence*, 12 March 2021, and 'China's stake in Africa's mines', *The Wire China*, 27 June 2021.

70   'Chinese Company removed as operator of cobalt mine in Congo', *The New York Times*, 28 February 2022.

71   'China's stake in Africa's mines', op. cit.

72   Watch the documentary *The Dark Side of Green Energies* (2020). Jean-Louis Pérez, Guillaume Pitron. France: Arte.

73   'China fires "warning salvo" in a direct threat to US; Can choke the world by blocking lithium, rare earth & API supply', *The EurAsian Times*, 6 July 2023.

74   'Battery "gigafactories" are flourishing across Europe', *Le Monde*, 15 May 2023.

75   See 'Tiamat veut mobiliser 100 millions pour sa première usine de batteries' ['Tiamat wants to raise 100 million for its first battery plant'], *Les Echos*, 19 January 2023. See also 'European manufacturing plan for sodium battery technology', *eeNews Europe*, 30 March 2021.

76   'IIT's rechargeable edible battery nominated on TIME's 2023 list of Best Innovations', IIT – Istituto Italiano Di Tecnologia, 24 October 2023.

77   'Water-activated paper battery named among world's best inventions', *Swissinfo.ch*, 11 November 2022.

78   'Crab and lobster shells could be used to make renewable batteries', *The Guardian*, 1 September 2022.

79   Read Vincent Donnen's fascinating paper 'Moving towards a metallic age: building industry resilience through a strategic storage mechanism for Rare Earth Metals', *Ifri Notes*, French Institute of International Relations, May 2022.

80   'US departments of Energy, State and Defense to launch effort to enhance national defense stockpile with critical minerals for clean energy technologies', Office of International Affairs, US Department of Energy, 25 February 2022.

81   See 'The Circularity Gap Report 2023', *Circularity Gap Reporting Initiative*, The Circle Economy Initiative.

82   The report forecasts that the annual consumption of metals alone should increase from 8 to 20 gigatonnes between 2011 and 2060. Refer to the 'Global Material Resources Outlook to 2060. Economic Drivers and Environmental Consequences', OECD, 12 February 2019.

83   Consult the European Parliament legislative resolution of 14 June 2023 on the proposal for a regulation of the European Parliament and of the Council on batteries and waste batteries, amending Directive 2008/98/EC and Regulation (EU) 2019/1020, and repealing Directive 2006/66/EC. Under the regulation, these figures are to be increased to 26 per cent cobalt, 12 per cent lithium, and 15 per cent nickel in 2036.

84   'The Circularity Gap Report 2023', op. cit.

85   'China to impose export controls on key materials for chipmaking as West's "chip war" escalates', *The Global Times*, 3 July 2023.

Chapter Nine: The last of the backwaters

1   'Montebourg veut que la France retrouve sa bonne mine' ['France Wants to Make a Mining Comeback'], *AFP*, 22 February 2014. The statement was made during the minister's visit to the gypsum quarries of Montmorency, north of Paris in 2014.

2   Appendix 21 shows a simplified map of mineral resources in mainland France, as identified by a mining inventory carried out between 1975 and 1992 by the French Geological Survey (BRGM), and updated several times up to 2021. See the report 'Evolution de`base de données 'gisements France': atlas des substances critiques et stratégiques' ['Update of the database of 'deposits in France': Atlas of critical and strategic substances'], BRGM/RP-71133-FR, December 2021. But experts agree that the inventory is obsolete, as it doesn't go beyond a subsoil depth of 300 metres.

3   'Macron enterre la Compagnie des mines de France chère à Montebourg' ['Macron Shelves Montebourg's Beloved National Company of French Mines'], *Challenges*, 9 February 2016. 'Montebourg dote l'Etat d'un bras armé dans le secteur des mines' ['Montebourg's Secret Weapon for France's Mining Sector'], *Les Echos,* 21 February 2014.

4   'Emmanuel Macron préside l'installation du groupe de travail chargé de définir la "mine responsable" du xxie siècle' ['Emmanuel Macron Presides over the Installation of a Working Group Tasked with the Definition of "Responsible Mining" in the Twenty-first Century'], Press Release from the French Ministry of the Economy and Finance, 1 April 2015.

5   '*Emmanuel Macron engage la démarche "mine responsible"'* ['Emmanuel Macron Launches the "Responsible Mining Initiative"'], *Minéral Info*, 28 March 2015.

6   The mining cadastre is available as an open source on data.gouv.fr. Detailed maps and a list of all mining permits and concessions since the nineteenth century can be consulted at camino.beta.gouv.fr. In 2022, Électricité de Strasbourg, a subsidiary of EDF, was granted exploration permits for lithium in the Bas-Rhin region. In the same year, Compagnie des Mines Arédiennes, a subsidiary of the Canadian company Aquitaine Gold Corporation, was granted licences to explore for a number of metals (gold, antimony, tantalum, copper, germanium, indium, platinum, molybdenum and lithium) at three sites south of Limoges. See 'De nouveaux permis de recherche d'or inquiète Stop Mines 87 dans le sud Haute-Vienne' ['New

gold exploration permits worry Stop Mines 87 in the south Haute-Vienne department'], *France Bleu*, 25 October 2022.

7   See 'Le premier étage de la réforme du code minier promulgué' ['Enactment of the first stage of the mining code reform'], *Minéral Info*, 22 August 2021, and 'Ordonnances du 13 avril 2022 complétant la réforme du code minier' [Decrees of 13 April 2022 complete the mining code reform'], *Vie publique*, 14 April 2022.

8   'The subsurface as a common good: a research programme for the responsible and sustainable use and exploitation of the subsurface', French Geological Survey (BRGM), 20 January 2023.

9   'Talga boosts Swedish battery graphite with new discovery', *TechInvest*, 9 October 2023.

10   'LKAB raises deposit estimate at Europe's biggest rare earth find', mining.com, 12 June 2023.

11   'Portugal's Lusorecursos gets environmental OK for lithium mine', *Reuters*, 8 September 2023.

12   'Vulcan Energy to mine 60% more German lithium than planned', mining. com, 13 February 2023. The mining industry is pinning a lot of hope on lithium extraction from geothermal water because of its small environmental footprint and low energy and water consumption of the technique.

13   'Creuser et forer, pour quoi faire? Réalités et fausses vérités du renouveau extractif en France' ['Digging and Drilling: what for? Realities and false truths about the revival of mining in France'], *Les Amis de la Terre France*, December 2016.

14   Ibid.

15   'Europe's green dilemma: Mining key minerals without destroying nature', *Politico*, 15 March 2023.

16   'Europe's largest lithium mine to open in France', *RFI*, 24 October 2022. According to the French Geological Survey, the deposit contains some 375,000 tonnes of lithium — enough to equip 700,000 electric vehicles with lithium-ion batters for the next 25 years, say Imerys. See 'Evolution de base de données 'gisements France': atlas des substances critiques et stratégiques' ['Update of the database of 'deposits in France': Atlas of critical and strategic substances'], op. cit.

17   'Une mine de lithium dans l'Allier' ['A lithium mine in the Allier region'], *France Nature Environnement Allier*, 14 December 2022.

18   In France, there are believed to be as many 3,500 former mines that are still polluted with heavy metals. See 'Mines: l'héritage empoisonné' ['The Poisoned Legacy of Mining'], *France Culture*, 5 May 2017. See also 'Exclusif: la liste des sites miniers empoisonnés que l'État dissimule' ['Exclusive: list of poisoned mining sites hidden by the State'], *Reporterre*, 4 June 2022.

19   'La ruée sur les métaux' ['The Metals Gold Rush'], *Le Monde*, 13 September 2016.

20   'Serbian PM sees no chance for reviving Rio Tinto lithium project', *Reuters*, 13 December 2023.

21   The exploration permit granted in 2013 was revoked in 2016 by Sweden's Supreme Administrative Court, citing an insufficiently detailed environmental

impact assessment. Local residents feared that nearby Lake Vättern, a source of drinking water for 300,000 people, would be polluted by heavy metal contamination. In 2021, Tasman Metals' application for a new mining lease was again refused. These administrative hurdles 'are scaring off a lot of many economic players looking to invest in Swedish mines', lamented Maria Suner, CEO of the Swedish Association for Mines, Mineral and Metal Producers. See 'EU battery push impeded by Sweden's mining hurdles', *Automotive News Europe*, 29 August 2022.

22 Les Amis de la Terre (Friends of the Earth) encourages recycling rare metals rather than moving back to mining. As attractive as this is, and given the meteoric growth in renewable energy, I do not believe their suggestion will put a dent in the growth of the mining sector.

23 'Emmanuel Macron préside l'installation du groupe de travail chargé de définir la "mine responsable" du xxie siècle' ['Emmanuel Macron Presides over the Installation of a Working Group Tasked with the Definition of "Responsible Mining" in the Twenty-first Century'], op. cit.

24 'Investir dans la France de 2030: remise au gouvernement du rapport Varin sur la sécurisation de l'approvisionnement en matières premières minérales et ouverture d'un appel à projets dédié ['Invest in the France of 2030: submission to the government of the Varin report on securing the provision of mineral raw materials, and on launching an appeal for projects to achieve this endeavour'], General Secretariat for Investment (SGPI), French Government website, 10 January 2022.

25 'Métaux stratégiques : les trois choses à faire pour sécuriser les approvisionnements en France (rapport Varin)' ['Strategic metals: three ways of securing supply in France (Varin report)'], *La Tribune*, 11 January 2022.

26 'Rare earth materials in the defense supply chain', US Government Accountability Office, 14 April 2010.

27 According to the report 'Metals for clean energy: Pathways to solving Europe's raw materials challenge', published in April 2022 by the Université catholique de Louvain and commissioned by Eurometaux (Belgium), producing one tonne of lithium on European soil could result in 71 per cent fewer CO_2 emissions than producing the same quantity outside Europe. This is on account of Europe's energy mix, which on average contains less carbon than in the rest of the world, particularly China, where most lithium is now refined.

28 'Russia owns the only plant in the world capable of reprocessing spent uranium', *Le Monde*, 29 November 2022.

29 'Déchets nucléaires: Greenpeace démonte les rails entre le Tricastin et la gare de Pierrelatte' ['Nuclear waste: Greenpeace dismantles the tracks between Tricastin and Pierrelatte railway station'], *Greenpeace*, 6 April 2010.

30 'Evolution statutaire et développement économique au menu des rois de Wallis-et-Futuna à Paris' ['Statutory changes and economic development on the menu for the kings of Wallis and Futuna in Paris'], *L'Express*, 11 July 2016.

31 There are in fact three kings (and three kingdoms) in Wallis and Futuna: two kings in Futuna and one king in Wallis. Following a royal dispute in 2017, only one of the two kings in Wallis is recognised in Paris.

32   Eufenio Takala came to power in 2016 and was still sovereign in 2023. Filipo
      Katoa resigned in 2018, citing health reasons. Lino Leleivai replaced him as ruler
      of the Kingdom of Alo, but stepped down in October 2022. A candidate has
      been put forward, but has yet been crowned, as the clans are at odds over how
      the ceremony should be conducted. See 'Royaume de Alo: un couronnement
      qui se fait attendre' ['Kingdom of Alo: A long overdue coronation'], *France Info*,
      10 May 2023.

33   The kings of Futuna are believed to be paid a much lower amount.

34   Interview with Pierre Simunek, former secretary-general of the prefecture of
      Wallis and Futuna, 2016.

35   G. Prinsen et al., 'Wallis and Futuna Have Never Been a Colony: A non-
      sovereign island territory negotiating primary education with metropolitan
      France', *Oceania*, vol. 92, no. 1, March 2022.

36   Bishop of Wallis and Futuna since 2019.

37   Interview with Pierre Simunek 2016.

38   French White Paper: Defence and National Security 2013, Ministère de la
      Défense, 2013.

39   Adopted in 2023, the French military programme law bolsters the resources of
      the French Navy in order to consolidate its protection of the French overseas
      territories. It represents €13 billion between 2024 and 2030. See 'LPM 2024–
      2030: les grandes orientations' [Military programming law: Key guidelines'],
      French Ministry of the Armed Forces, April 2023.

40   See 'Mission IRD. Présentation' [Mission of the French Research Institute
      for Development. Presentation'], Wallis and Futuna, *IRD*, 7-17 September 2018
      (wallis-et-futuna.gouv.fr).

41   Interview with Pierre Simunek, 2016.

42   'Wallis et Futuna opposés à toute forme d'exploration minière' ['Wallis and
      Futura opposed to any form of mining exploration'], *France Info*, 19 September
      2018.

43   See 'L'île de la Passion-Clipperton, aux confins géopolitiques de l'Indo-
      Pacifique français' ['Clipperton Island: the further reaches of the French Indo-
      Pacific'], *The Conversation*, April 2023.

44   See 'La ruée vers les grands fonds s'accélère' ['The scramble to the bottom of
      the sea gathers pace'], *Le Temps*, 30 April 2019 and 10 June 2023, and 'Deep-sea
      minerals could meet the demands of battery supply chains – but should they?',
      mining.com, 31 July 2020. Japan also discovered 16 million tonnes of rare earths
      in the Pacific Ocean near Minamitori Island in 2018. Enough to 'meet centuries
      of global demand', according to Japanese researchers. See 'Des quantités "quasi-
      infinies" de terres rares dans les eaux japonaises?' ['Quasi-infinite quantities of
      rare earths in Japanese waters?'], *Usbek et Rica*, 12 April 2018.

45   See 'L'exploration et l'exploitation des fonds marins' ['Deep-sea exploration
      and exploitation'], Comparative Legislation Study No. 305, French Senate, July
      2022, and 'La course pour l'exploitation des ressources du plancher océanique
      s'organise' ['The race to mine resources on the ocean floor heats up'], *Le Monde*,
      21 October 2021.

46   In March 2023, Loke Marine Minerals acquired UK Seabed Resources from
      US defence group Lockheed Martin. See 'Lockheed Martin sells deep-sea
      mining firm to Norway's Loke', *Reuters*, 16 March 2023.

47   Interview with Pierre Cochonat, currently a consultant in marine geosciences,
      2013.

48   France has set up the Extraplac programme (Extension raisonnée du plateau
      continental — www.extraplac.fr) dedicated to requests for extensions beyond
      the exclusive economic zones.

49   'Commission on Limits of Continental Shelf Concludes Fifty-Seventh
      Meeting', Commission on the Limits of the Continental Shelf, United
      Nations (UN), 16 March 2023.

50   A practice repeatedly condemned for years and a source of tension with
      neighbouring countries (Vietnam, Malaysia and the Philippines). See, for
      example, 'La Chine accusée de bâtir des îles artificielles pour étendre sa zone
      maritime' ['China accused of building artificial islands to extend its maritime
      zone'], *Les Echos*, 28 December 2022.

51   Max Mönch, Alexander Lahl, *Ocean's Monopoly*, produced by Werwiewas, 2015.

52   'A climate solution lies deep under the ocean. But accessing it could have huge
      environmental costs', *TIME Magazine*, 13–20 September 2021.

53   'Minage en eaux profondes: On ne peut pas parler d'exploitation durable,
      de toute façon' ['Deep sea mining: sustainable mining does not exist'], *RFI*,
      8 February 2022.

54   COP 27 was held in November 2022 in the Egyptian city of Sharm el-Sheikh.
      See 'Emmanuel Macron veut interdire l'exploitation des fonds marins'
      ['Emmanel Macron wants to ban deep sea mining'], *Le Monde*, 7 November
      2022. A few months later, the French National Assembly also voted against
      deep-sea mining by an overwhelming majority. See 'L'Assemblée nationale
      se prononce contre l'exploitation minière des fonds marins' ['The National
      Assembly votes against deep sea mining'], *Le Monde*, 17 January 2023.

55   'Report on international ocean governance: an agenda for the future of our
      oceans in the context of the 2030 SDGs', Report A8-0399/2017, European
      Parliament, 18 December 2017.

56   'Des parlementaires du monde entier demandent un moratoire sur
      l'exploitation minière des grands fonds marins' ['Parliamentarians around the
      world call for a moratorium on deep-sea mining'], *Europe Ecologie*, 29 June 2022.
      Moreover, in October 2022 the European Investment Bank (EIB) defined the
      extraction of mineral deposits from the deep sea as 'unacceptable in climate
      and environmental terms', and, as such, will not finance it. See 'EIB Eligibility,
      Excluded Activities and Excluded sectors list', European Investment Bank,
      2022.

57   Watch the documentary *Ocean's Monopoly*, op. cit.

58   H.R.2262 — US Commercial Space Launch Competitiveness Act, 114th
      Congress (2015–2016).

59   See 'Asteroid with platinum core worth £3.5 trillion set to pass Earth',
      The Independent, 18 July 2015.

60　'NASA is set to explore a massive metal asteroid called 'Psyche' that's worth way more than our global economy', *Forbes*, 28 December 2021.

61　'Commercial asteroid mining now has a 2023 launch date to scout its first target', *Forbes*, 24 January 2023, and 'Space mining startups see a rich future on asteroids and the moon', Space.com, 7 January 2023.

62　'Who owns the moon?', *Deutsche Welle*, 15 November 2022.

63　'Will a five-year mission by COPUOS produce a new international governance instrument for outer space resources?', *The Space Review*, 10 February 2023. Work began in late 2022.

64　Read the fascinating article written by the astronaut Thomas Pesquet, 'Mines dans l'espace, la nouvelle frontière' ['Mines in Space: the new frontier'], *Les Échos*, 8 October 2017.

65　Luxembourg is hardly a novice, as it already headquarters the world's leading satellite operator SES Group.

Epilogue

1　Michio Kaku, *Physics of the Future: how science will shape human destiny and our daily lives by the year 2100*, Allen Lane, 2011. See also 'China to Build World's First Solar Power Station in Space in Next Five Years', *Newsweek*, 4 November 2019.

2　'Photovoltaïque: les promesses des pérovskites' ['Photovoltaics: the promises of perovskites'], *Le Monde*, 15 June 2017.

3　See Pierre-Noël Giraud and Timothée Ollivier, *Économie des matières premières* [*The Economics of Raw Materials*], op. cit.

4　See Philippe Bihouix, *L'Âge des low technology: vers une civilisation techniquement soutenable* [*The Age of Low Technology: towards a sustainable technical civilisation*], Seuil, 2014.

5　Translator's note: Originally a French term (*décroissance*), 'degrowth', refers to the downscaling of production and consumption (energy and resources), and to decoupling growth from improvement.

6　Interview with Christian Thomas, 2017.